COLD WAR TRIANGL

HOW SCIENTISTS IN EAST AND WE

Cold War Triangle

How scientists in East and West tamed HIV

Renilde Loeckx

Lipsius Leuven

Lipsius Leuven is an imprint of Leuven University Press.

ISBN 978 94 6270 113 7
D/ 2017 / 1869 / 31
NUR: 694

Book design: Dogma

To Bill with love

Contents

Foreword

It was in a small town in the Czech Republic that I first met the scientists who moved the fight against viral disease a giant leap forward. As Ambassador of Belgium, I was invited to attend the ceremony at the University of South Bohemia conferring honorary doctorates to the late Antonín Holý and Erik De Clercq. Their cooperation with scientists of an American company, Gilead Sciences, was at the origin of a series of miracle drugs which are the most widely used drugs today, not only to combat AIDS but, to actually prevent HIV infection. It struck me how little the general public knows about the scientists who saved millions of lives and will safeguard millions more in the future. Although I am not a scientist but a retired diplomat, I felt compelled to tell their story.

In my former life, I was better skilled at observing the corridors of political power and organizing cocktail parties than probing the exact methodical world of science. Embarking on this venture, I felt I needed to bridge a gap between two totally different cultures, one way of thinking a mile wide but an inch deep, the other an inch wide but a mile deep.

How to overcome these obstacles in communication? Some of my previous encounters with scientists during my diplomatic career gave me confidence.[1] My meeting in Philadelphia with Renée Fox, Professor at the University of Pennsylvania emboldened me most. She had observed the many talented young European physicians at Harvard Medical School just after World War II preparing for research careers in academic medicine, and wondered what sort of conditions and problems they encountered in their home settings. Belgian medical research fascinated her, it became her favorite subject of study for over thirty years.[2] The fact that so many cultural influences were concentrated in a country no larger than the state of New Jersey had piqued her curiosity. Her writings and her words have inspired me throughout this journey.

Cold War Triangle looks beyond Belgian medical research, and covers academic institutions in other countries, particularly in former Czechoslovakia, and how their research was combined with the genius of

American science and entrepreneurship. It not only straddles the workings of scientists across oceans and continents but also across deep political and ideological divides.

My main source of encouragement in the writing of this book was undoubtedly my American husband, William (Bill) Drozdiak. As a foreign correspondent and later as editor of The Washington Post, he taught me all along our parallel careers to look out for the so-called "nugget," what people are really trying to say. The all-important story they bury under hints and insinuations or in messages that people cry out at the top of their lungs but that nobody hears. This book is dedicated to him.

Acknowledgements

I am immensely grateful to Erik De Clercq, Professor Emeritus at KU Leuven. His gift of friendship made this book possible. As one of the pioneers in antiviral drugs he patiently coached me through the universe of chemists and virologists. I greatly benefited from his knack of teaching and instinct of keeping co-workers and students alike focused on a common goal. His mantra "Keep your eyes on the prize, we get what we focus on" infected me with the "virus" to dig further and sharpen my questions.

His once closest collaborators, Professor Piet Herdewijn, Professor Johan Neyts, Professor Dominique Schols, Dr. Rudi Pauwels founder and CEO of Biocartis, and Professor Emeritus Jan Desmyter helped me understand the scope of the accomplishments they achieved as a team. Professor Emeritus Alfons Billiau imparted precious background knowledge on the origins and workings of the Rega Institute of Medical Research. My special thanks and warm appreciation also to Christiane Callebaut, she was my tireless go-between and precious contact at the Institute.

Travels to the Czech Republic, Warsaw, New York, Paris and California opened the doors of Erik De Clercq's wide network of friends and colleagues. I felt very fortunate to witness the 100th birthday in Warsaw of the late Professor David Shugar. He was one of Erik De Clercq's first co-workers who introduced him to the world of nucleosides.

I am deeply indebted to Professor Zdenek Havlas, former Director of the Institute of Organic Chemistry and Biochemistry in Prague and Professor Libor Grubhoffer, former President of the University of South Bohemia. They invited me to several key events in the Czech Republic honoring the late Antonín Holý and spared no effort to guide me through the Czech labyrinth. The reminiscences of Dr. Yvan Rosenberg and Dr. Radim Nencka gave me unique insights into Prof. Antonín Holý's accomplishments. His widow, Ludmila Holá, and his daughter, Dr. Dana Holá, enlightened me on some of his most endearing habits. I very warmly thank both Professor Jan Vilcek and Professor Marc Van Montagu, who gave me lively accounts of the lives of scientists in the former Czechoslovakia.

I am particularly grateful to Dr. John Martin, Executive Chairman of Gilead Sciences. He was the person who held the key to leading this multi-national, multi-lingual team to success. His *band of brothers* who were his faithful companions from the very beginning of this venture in Foster City, California, put the pieces of the puzzle in place: Dr. William (Bill) Lee, Dr. Norbert Bischofberger, Dr. Mick Hitchcock, Dr. Swami Swaminathan generously gave me their perspective on how the miracle drugs came into being.

My special thanks go to Michael Riordan, the founder of Gilead Sciences and former CEO, who took me back in time to the early days of the company, and to Professor Richard Whitley, Director of the Gilead Board, who graciously recounted the quest to find drugs against viral diseases and the crusades they embarked on to convince the medical community and policymakers. I was very fortunate to meet Dr. Gregg Alton of Gilead Sciences, who made me understand the company's unique procedures and innovative ways of bringing their drugs to the people most in need, to Africa and beyond.

I greatly valued the guidance of Professor Emeritus Peter Piot, Dr. Lori Lehman and Ms. Amy Flood from Gilead sciences, Mrs. Marie-Anne De Somer, Dr. Robert Redfield from the University of Maryland in Baltimore, Dr. Richard Ennis, Frm. Secretary General of the NATO-Assembly Simon Lunn and Ms. Judith Miller. Each helped me find a path through the many angles of this story. A special appreciation is due to Dr. Vladimir Beroun who taught me how to master my electronic devices.

Among my cheerleaders, I would like to single out the enthusiastic support from Hugo and Lucie Van Geet as well as Jaroslav Kurfürst, the Ambassador of the Czech Republic in Brussels.

I owe a major debt of gratitude to my children, Karen Drozdiak and Nicholas Drozdiak, my line editors, who cheerfully reminded me that English is not my native language. My youngest daughter, Natalia Drozdiak, wrote the booklet *The Virus Hunters: Banishing the Scourge of AIDS*. It became a calling card for the virology department of the Rega Institute. It stimulated the interest and enthusiasm of the editors of Leuven University Press, Veerle De Laet and Viktor Edmonds, who patiently guided me throughout my effort to write on a subject beyond my previous scope of expertise.

ACKNOWLEDGEMENTS

The author's proceeds of this book will be donated to the European Organization for Research and Treatment of Cancer. EORTC manages the research of more than 5,000 oncologists and cancer centers in 870 university hospitals in a Europe-wide network.

Introduction

Excellence is rarely found, more rarely valued.
— Johann Wolfgang von Goethe

"Cold War Triangle" is about the human face of science, how scientists from three different cultures collaborated to create the complex drugs that saved millions of lives. Who were the mentors that influenced them and their careers? How did they intersect with one another? What traits in their unique backgrounds and disparate journeys led them toward the making of key discoveries in modern medicine? This book recounts an inspiring story of the groundbreaking cooperation between East and West during the darkest days of the Cold War. How did scientists behind the Iron Curtain overcome authoritarian rule, cross hostile borders and ultimately collaborate with colleagues in the West? Who would have thought then that their cooperative spirit would culminate in a vital weapon to inhibit HIV and thwart an epidemic?

Preventing the onslaught of infectious disease has been an innate human concern since the dawn of time. The discovery of antibiotics to combat bacteria laid the groundwork for a new medical field, virology, which focused on the tiniest of microbes. The development of vaccines to prevent debilitating diseases and save lives brought relief to mankind. The search for antiviral drugs was initially considered extraneous. Unaware of a looming epidemic, some scientists were already starting to put the tools in place to combat a retrovirus.

HIV is a slow-moving retrovirus but contrary to all other viruses-causing infectious disease, it has proven to be one of the deadliest forms that humanity has ever encountered. If not treated, this virus kills almost all without fail. When HIV came into the limelight in the 1980s, people became painfully aware that there were hardly any antiviral drugs in existence. None of them could inhibit HIV. Even the entire arsenal of antibiotics, one of the biggest triumphs of medicine, proved useless against viruses.

AZT, the first drug to treat AIDS, became available in 1987. However, it gave only a short reprieve as it proved too toxic for long-term treatment. Death continued to lurk at the doorstep of those infected with HIV. When science could not produce a real life-saving treatment, disillusion turned into anger. The streets exploded with demonstrations by gay protesters, besieging large pharmaceutical companies. As the AIDS epidemic expanded and more lives were lost, the media and public demanded to know why modern science could not find an effective treatment.

It took almost ten more years until a new generation of drugs hit the market. Patients struggled to keep pace with their therapy; the mix of medications involved taking more than twenty pills at six different times of the day and still had plenty of side effects. It kept people alive but in a miserable way.[3]

A triangular partnership, formed during the Cold War between Belgian, Czech and American scientific teams, led the way to the most effective treatment in the world today that features a one-a-day pill with few side effects. The protagonists—Erik De Clercq, a Belgian scientist from the University of Leuven, and his Czech colleague, Antonín Holý, from the Academy of Sciences in Prague—first met in West Germany in 1976. Their work yielded splendid discoveries that were licensed to an American company before the fall of the Berlin Wall. Holý and De Clercq had single-handedly created a whole class of molecules out of which *Tenofovir* emerged. The conviction of one man, John Martin, was the driving force behind his team at Gilead Sciences to develop this compound. Tenofovir has since evolved into several drugs that allow HIV-infected people to lead a normal life. One of them, Truvada, has also been approved for the prevention of HIV, which could put an end to the epidemic if enough people could take it. The Gilead drugs became the gold standard for HIV treatment in the West. John Martin played a pivotal role in bringing these twenty-first century drugs to Africa and the rest of the developing world.

A brief excursion into the history of virology and vaccines

The power of a microbe is a baffling phenomenon. It can unseat empires and shape history. The end of the Roman Empire was connected to a steep demographic decline caused by smallpox epidemics, which had been spread throughout the Mediterranean by Roman soldiers. The bacteria causing Black Death (Yersinia pestis), which originated in China, was carried on the backs of Mongolian hordes to a major Genoese trading port on the Black Sea. The bacteria travelled on black rats in cargo destined for Europe.[4] The Black Death that swept the continent in the fourteenth century unseated the dominance of Christianity and sparked the Renaissance.

In 1492, Christopher Columbus's ships introduced a host of new microbes to the New World, which unleashed diseases onto the Native American population. A combination of smallpox and measles caused a devastating drop in population. After those diseases weakened the Aztec capital in 1519, it allowed for a small Spanish army to build alliances and conquer the empire. Twenty years later, an epidemic, most probably smallpox, similarly aided the Spanish troops when they invaded the Inca Empire.

People believed these diseases were caused by the rare conjunction of planets, miasmic vapours or the wrath of the gods. They did not realize that tiny microbes were in fact the culprits.

Microbiology emerged in the 19th century, when French wine and beer chemist, Louis Pasteur, disproved the spontaneous generation of living things. By using filters to exclude dust particles, he could prevent mould from growing in boiled broth. Pasteur was the one who alerted the world to the deadly power of microbes. He knew that some of these germs could be useful in wine and beer making but he had also observed how physicians knew less about antiseptic methods than vintners or brewers did. He accused medical doctors of causing the death of women during childbirth as they transmitted and infected them with microbes from other patients.[5]

His rival in Berlin, Robert Koch, established strict criteria, now known as the Koch postulates, for linking a microbe to a disease. In the 1870s, any infectious invader, bacterial or otherwise, was considered a virus, derived from the Latin word for poison. Only the larger microbes like bacteria could be seen under microscopes.[6] The tiniest among the microbes, the viruses, remained invisible and their existence could only be surmised.

Louis Pasteur became famous for treating the deadly rabies virus after noticing there was an infectious substance so small it could pass through his filters but could not be seen under a microscope.[7] He devised a vaccine by growing microbes in living animals and using them as a weapon against their own kind. His injections saved a boy bitten by a rabid dog. They won the race against the slow-moving but deadly virus before it had completed its incubation time. He singlehandedly put an end to the savage killings of humans from packs of rabid dogs that used to roam the French countryside.

This was not mankind's first vaccine. Throughout medical history there was an intuitive knowledge that survivors of some infections like smallpox became immune to the disease. Calculated exposure to infectious disease in order to stimulate immunity was known in many parts of the world.[8]

Vaccines were first discovered in the late eighteenth century after country doctors in rural England noticed that dairymaids with small lesions had become infected by the harmless cowpox and yet were fully protected against smallpox. One of these doctors, Edward Jenner, tried to use that cowpox, much weaker than the virulent smallpox virus, for inoculations in humans, and discovered he could protect people from the dangerous disease. He proved that a foreign substance similar to but not as dangerous as the one that causes disease could trick the body into believing it was under attack and stimulate it into making antibodies. The *vaccinia virus*, the Latin name for cowpox virus, became the operative word in "vaccination."[9]

It took almost eighty-five years from Jenner's time until Pasteur developed mankind's second vaccine. Pasteur realized that the body's memory of its encounters with microbes was key to its defense. It led to the building of antibodies and resistance to prevent re-infection by the same microbe. At the turn of the century, Paul Ehrlich studied this phenomenon more closely and called it "immunity." During the first half of the twentieth century, scientists added five more vaccines to save mankind from crippling and deadly diseases. The vaccines against diphtheria, tetanus, and whooping cough were cultured out of bacteria mixed with chemicals.[10]

Knowledge about viruses remained in the dark for a long time. Even once the electron microscope was invented in the 1930s, viruses were often disregarded as chemical elements, not living organisms. It was not until the late 1940s that the modern concept of a virus emerged. Viruses resemble seeds; they can only spring to life when they find the right soil. They must find a cell to infect in order to survive. Only when they have entered their host's cell can their genetic material be reproduced with new viruses ready to infect more cells.[11]

The true revolution that boosted the growth of virology came in the late 1940s with the development of cell culture techniques by the American scientist John Enders. Animal or human cells could now be grown in laboratory flasks and would indicate, through a change in the appearance of the cells, the presence of a harmful virus. Before that time, researchers had to resort to live animals or chopped up animal organs in order to detect and identify viruses. Not surprisingly, only eight viruses dangerous to man had been found in the first half of the twentieth century and nobody was able to coax them to reproduce inside a test tube.[12] With the new techniques, a torrent of human disease-causing viruses could be isolated and the science of virology progressed exponentially.[13]

Chapter I
Leuven: a hotbed for antiviral research

A true fundamental researcher is an introvert who takes pleasure in looking for answers to questions that nobody asked.

— Piet De Somer

The cross-fertilization between academia and pharma

In the early years of World War II, the small university town of Leuven was suffering badly from the Nazi occupation. The Germans had again ransacked its world-famous library, which had been devastated during the First World War. In a drafty old building, a young researcher named Piet De Somer and his boss were studying the behavior of a strain of penicillium they had smuggled from the Netherlands.[1] They were fascinated by reports that it could produce a new infection-fighting drug.

British war broadcasts and Swiss medical journals had revealed that American companies were producing a miracle drug based on a discovery by the Pathology School of Oxford University.[2] Unlike their British counterparts, the Americans had sensed the strategic importance of this discovery and alerted the Roosevelt Administration. In 1941, the production of penicillin became part of an urgent government-industry venture with the sole purpose of making the drug available to the troops so that soldiers would not perish from infectious diseases.

Producing penicillin seemed simple enough; it required cultivating an omnipresent penicillin mold similar to the one that had accidentally contaminated Alexander Fleming's bacterial culture.[3] The British discovery as such was not patented. The technical protocol on how to mass-produce and extract the penicillin from the culture fluid, however, was guardedly protected by the American pharmaceutical industry. Secrecy surrounded the penicillin production even after the war was over. Hospitals and doctors in the rest of the world literally begged the Americans to obtain a few ounces of penicillin. Producing this drug on Belgian soil would become a matter of national pride.[4]

Piet De Somer's boss and two fellow professors at the Catholic University were eager to take up the challenge. They were partners in a small company, *Soprolac*, an offshoot of a cheese company that doubled up as a pharmaceutical business. The byproduct of the cheese-making was used to produce *Panferma*, a medicinal water to treat all kinds of aches and pains. After *Soprolac* was purchased by a young Belgian industrialist active in the paper industry, the academic co-owners suddenly became partners in a much larger enterprise named RIT (*Recherches et Industries Thérapeutiques).*[5] Piet De Somer was entrusted with their goal to produce penicillin. But he had one problem. His knowledge of chemistry was modest and purifying the product after he had cultured the mold was a complicated process.

That is when his legendary charm came to the rescue. He befriended a fellow medical student, Christian de Duve, who was working on a Master's degree in chemistry and needed a topic for his thesis. At that time, there were shortages in the lab, so they took discarded milk bottles from the former *Soprolac* plant to culture molds.[6] De Somer and de Duve shuttled daily between Leuven and Genval to monitor their cultures. Communication and travelling were still very restrained in those days, but their trips were quite flamboyant. De Somer drove an *Amilcar 1928*, a racecar which had neither roof nor battery and needed a roller bearing crankshaft to jumpstart the car. Wherever they went, they were greeted with roaring laughter.[7] De Duve succeeded with the purification and thus the first milligrams of penicillin were produced on Belgian soil. He eventually moved on to other research projects and later won the Nobel Prize in 1974, but he always kept fond memories of those wild times.

The RIT co-owners realized that in order to produce larger yields of penicillin a "deep fermentation" method would be needed instead of the artisanal "surface culturing." The American companies were not sharing information. However, the professors found a way to circumvent them. They had excellent contacts with the Director of Connaught Medical Research Laboratories in Toronto who had just started his own production of penicillin.[8] In 1946, they dispatched Piet De Somer to Toronto where he was introduced to in-depth culture production of penicillin.[9] He returned to Belgium a few months later and set up a small fermentation plant for RIT. Together with his co-workers, he spent the next few years working intensively on improving the mass-scale production of penicillin. Not encumbered by any license, it became a huge financial success and gave De Somer the incentive to look for other antibiotics.

Penicillin proved very effective against some bacterial infections such as those that caused blood poisoning. But it was useless against other bacteria that caused such mortal illnesses as tuberculosis, cholera, or urinary and intestinal infections.[10] Tuberculosis, a scourge known throughout history as the White Plague, had been contained in recent years mainly thanks to better sanitation but it remained a major public health issue due to its contagious nature.

While stories about another miracle drug, streptomycin, coming from Selman Waksman's lab at Rutgers University spread like wildfire, research with streptomyces molds in the antiquated Leuven laboratory had been less than successful.[11] RIT, however, was eager to acquire streptomycin to fight tuberculosis and a host of other bacteria.[12] It was left to Piet De Somer to negotiate with Waksman and purchase a production license. He also acquired the license for the production of aureomycin. This helped to transform the small RIT facility into a well-oiled manufacturing plant that reaped huge financial successes.[13]

The time had come to create a modern research facility in Leuven.[14] An agreement was reached whereby the university would provide the land and Piet De Somer's industrial partner would erect the research building. The company's research laboratories would become part of a new Institute. Its structure, bridging sections from two rival university faculties, medicine and pharmacy, required uncommon skills. De Somer, Head of the Institute and at the same time Director of the university department of microbiology, possessed the charm and wiles of a Florentine prince.[15]

The Director of the university hospital was given the privilege to name the institute. The choice was *Rega* after Hendrik Joseph Rega, a renowned scholar of the 1700s and author of several medicinal treatises in Latin.[16] The name *Rega* was an auspicious omen for close cooperation between academia and industry, which was an entirely new phenomenon in Europe in 1954. It would also marry medicinal chemistry with microbiology, a new virology branch that was an extravagant novelty for the University of Leuven at the time.[17]

When the new Institute opened its doors, virology would become the heart of its work. Piet De Somer sent his assistants to other virology laboratories in Europe where they studied the equipment and copied protocols necessary to tackle this new science in Leuven.[18] He closely followed all these exciting developments, especially the cell culture techniques and sensed the imminent explosion in vaccine research. He wanted his group to become the first and the best in the field.[19]

Vaccines and celebrity scientists

In the Fifties, commotion over paralytic polio engulfed the United States. Despite health statistics in the years after World War II concluding that children were three times more likely to die of cancer and ten times more likely to be killed in car accidents than by polio infections, polio kicked up a media storm and gained special status as a public scourge that required urgent treatment. This was due in large part to the efforts of the National Foundation for Infantile Paralysis, which employed the latest techniques in advertising, fund raising, and motivational research to transform a relatively uncommon disease into the most feared affliction of its time.

The media hype spread to Belgium and other countries across Europe.[20] Polio was not the deadliest of viruses, but it was an insidious one. Less than one in a hundred of those infected showed symptoms of paralysis.[21] It was precisely those infected who did not show any symptoms that caused the virus to spread. Those who were struck by paralysis were handicapped for life.

In the United States, panic often escalated into mass hysteria. Every time a polio wave emerged, usually during the summer months, swimming pools and movie theaters would close, beaches and streets were deserted and spraying of DDT or other insecticides was used on a massive scale to sanitize cities, and sometimes even the interiors of houses.[22] Nobody knew then that the disease was actually an outgrowth of improved sanitation and that those who lived in cleaner, more comfortable homes were at greater risk.[23] Much later it became clear that young children who lived in crowded and unsanitary conditions, such as households without indoor plumbing or toilets, had developed resistance by exposure to the poliovirus at a young age when they still benefited from their mother's antibodies. The virus had already been isolated in 1908, but it took almost half a century before a preventive remedy was introduced, a vaccine based on the killed virus.[24]

The inventor and savior, Jonas Salk, was born and raised in New York City[25]. He had discovered that there were three different polio strains and that antibodies against one did not offer protection against infection from another. Salk grew the viruses in cell cultures following the new discoveries of John Enders and subsequently killed the viruses without destroying their immunizing power.[26] His killed-virus vaccine could trick the immune system into believing that the body was under attack and needed to react with an increase in protective antibodies. On 12 April 1955, Jonas Salk's polio vaccine had formally been declared safe, effective and potent. He was hailed a hero. To many in the US, April 12 resembled another V-Day, the end of a war. People huddled around radios to hear the news, some wept openly with relief, outside one could hear car horns honking and church bells chiming in celebration. President Eisenhower declared Salk a benefactor of mankind and honored him with a Rose Garden ceremony.[27]

Piet De Somer met Jonas Salk during a conference in Stockholm and began corresponding extensively with the celebrity scientist to discuss better ways to deactivate and filter the virus.[28] He convinced the Board of Directors of the RIT company to start development of the polio vaccine. Production was to be launched in the Rega Institute in 1955, the year that De Somer was appointed as a full-fledged Professor in the medical faculty. A whole colony of monkeys was promptly housed on the top floor of one of the Rega buildings, as monkey kidney cells were needed to culture the poliovirus.[29]

In those days, as recounted by one of De Somer's assistants, culture media had to be "homemade." Growth-promoting serum had to be obtained through the centrifugation of blood collected from horses or calves in local slaughterhouses. Personnel were still inexperienced and not yet accustomed to antiseptic techniques. It was a constant struggle to combat impurities in cultured cells. Moreover, the process involved the complicated acquisition of the three types of virulent polio strains.[30] The new Institute overcame all these hurdles.

The polio strains purchased from the Institut Pasteur arrived in Leuven in November 1955. A few months later, the vaccines were ready. They were first administered in the Rega Institute, starting, as dictated by tradition, with Piet De Somer and his children. Unhindered by rules and regulations, the vaccine was distributed in some schools later that year. Soon, the polio vaccine operation was moved to the RIT company facility in Genval-Rixensart for large-scale production. To aid the testing of RIT production, one of De Somer's newly hired assistants was sent to Pittsburgh to work in Salk's lab as a visiting scientist.[31] The Dutch and German governments were among the first international buyers of RIT vaccines. The Swedes, Danes and French had all been involved in vaccine production several years before the first cell cultures were established in Leuven. However, in less than 24 months, Rega and RIT had caught up with their competition in Europe.[32]

In 1958, at the height of the media frenzy around polio, the first postwar World Fair was held in Brussels.[33] For Piet De Somer, it was a unique opportunity to showcase Belgium as one of the first countries outside the US that could produce its own polio vaccines. De Somer's reputation as a scientist-entrepreneur achieved iconic status.

Meanwhile Salk's rivals were working on an innovative live virus vaccine. Instead of killing a virulent virus, they used a living virus that is non-virulent to begin with and weakened it in animal cells.[34] Piet De Somer kept a close eye on his competitors and especially on Albert Sabin, who was conducting the largest trials in medical history for his vaccine in the Soviet Union.[35]

His assistant met with Albert Sabin during a conference in Tokyo in September 1961 and a few months later, the live attenuated virus vaccine was manufactured by RIT in Belgium. The speed with which the Rega Institute and RIT went into production astonished everyone. Only a few months after Albert Sabin's oral vaccine replaced Salk's injectable vaccine in the USA in 1962, it was introduced in Belgium. By 1967, Belgium would become one of the first polio-free countries.

The live virus vaccine, usually disguised in a cherry tasting syrup or wrapped in candy, became the vaccine of choice in most parts of the world.[36] As to Albert Sabin himself, he was more popular outside the US where Jonas Salk would remain forever the beloved hero. In Europe, and the Soviet Union in particular, Sabin acquired the celebrity status he could not achieve at home.

Chapter II
Behind the Iron Curtain

Professionally, scientists and mathematicians are strictly international-minded and guard carefully against any unfriendly measures taken against their colleagues living in hostile foreign countries.

— Albert Einstein

The molecular revolution goes east

Throughout the early Fifties, McCarthyism dominated life in America and held all of Europe in its grip. Republican Senator Joe McCarthy had created a poisonous climate of fear and suspicion. His communist witch-hunting pushed the FBI into spying on citizens, thousands were accused of being communists or communist sympathizers and became the subject of aggressive investigations and questioning before government or private-industry panels, committees and agencies. The atmosphere of suspicion sapped the energies of scientists like Linus Pauling and Robert Oppenheimer among others. Even Jonas Salk was interrogated about the communist sympathies he once harbored in his youth.

It was in this dark period that one of the most remarkable collaborations between East and West took root. In July 1955, Bertrand Russell, the British mathematician, philosopher, and writer, issued the now famous Russell-Einstein Manifesto, cosigned by seven other Nobel laureates. He strongly believed only dialogue could avert a catastrophic thermonuclear war:

> In the tragic situation which confronts humanity we feel scientists should assemble in conference to appraise the perils that have arisen as a result of the development of weapons of mass destruction, and to discuss a resolution in the spirit of the appended draft. We are speaking on this occasion, not as members of this or that nation, continent or creed but as human beings, members of the species Man, whose continued

> existence is in doubt. The world is full of conflict; and, overshadowing all minor conflicts, the titanic struggle between Communism and anti-Communism.

Russell was the prime mover behind conferences bringing scientists from East and West together. Cyrus Eaton, the American railroad mogul, was the financier who generously invited the scientists to convene in idyllic places. The first meetings with mostly physicists, and only a few chemists and biologists, took place in the small village of Pugwash, Nova Scotia, where Eaton had grown up. There were no hotels to accommodate the conference participants in Pugwash so Eaton transformed a few of his sleeper trains into a makeshift hotel. Nobody would have guessed from these modest beginnings that the conferences would become such a powerful undercurrent in an era of McCarthyism and Cold War tensions. The Pugwash conferences gave scientists the backbone to cultivate cooperation between East and West.

During the coldest periods of the Cold War, when diplomats distrusted each other most, *Pugwash* successfully established a climate of trust among influential scientists from the East and West. Every time Cold War tensions increased, the role of the informal East-West backchannel of contacts between scientists became more crucial.[1] Although the annual conferences were condemned in the US Senate, both Eisenhower and Khrushchev came to appreciate them.[2] Following a *Pugwash* held in Moscow, scientists, not diplomats, laid the groundwork for a major diplomatic achievement, brokering the first major pact between East and West to limit nuclear tests—the Limited Nuclear Test Ban Treaty.[3] Yet to this day, the Pugwash movement remains largely unknown to the general public.

In 1955 during a Rose Garden ceremony honoring Jonas Salk, President Eisenhower surprised the world when he announced that "he would give the Salk vaccine formula to every country that welcomed the knowledge, including the Soviet Union."[4] It didn't take long for Soviet scientists to arrive in Washington following Eisenhower's generous invitation to "study polio and the preparation of the Salk vaccine."

Still reeling from the wounds inflicted by the McCarthy investigators, Jonas Salk declined to cooperate with his Soviet counterparts. Albert Sabin, however, jumped on the occasion. His live virus vaccine was ready but had not yet been tested on human beings. His work with a topflight virologist heading the Soviet Polio Research Institute led to the Soviet decision in 1959 to use Sabin's cherry flavored liquid for the oral vaccination of more than ten million children. The trials were conducted with a discipline akin to a military campaign and became the largest field trials ever in medical history.[5] The coercive powers of a police state were certainly part of its success but the field tests were also a signal that cooperation between East and West was not impossible.

The head of the Polio Research Institute, Mikhail Chumakov, and his wife were part of the movement to liberate Russian science from Stalinist quackery. Since the 1930s, Stalin's favorite scientist, Trofim Lysenko, had dominated Russian biology. Lysenko faked and tampered his data to make them consistent with the broader Marxist doctrine and show that characteristics acquired through life under socialism were more important than hereditary transfer of stable genetic properties. His notoriety dated back to a time of widespread famines caused by the forced collectivization of agriculture. Stalin's support propelled his theories from agriculture into human genetics. It became one of the biggest scientific scams in history and proved a huge setback for Soviet scientists.

Lysenko vehemently opposed the fact that organisms contained minute particles known as genes that are responsible for the transmission of hereditary traits. He dismissed this discovery as bourgeois pseudo-science. His main target was the father of genetics, Gregor Mendel, and his famous experiments in the garden of the Augustinian monastery at Brno in 1865 when Czech lands were still part of the Austro-Hungarian empire.

Lysenko strongly refuted the fact that DNA holds the key to genetic information in cells.[6] Soviet geneticists who rejected "Lysenkoism," and propagated DNA as the carrier of genetic information, faced public denunciation, loss of employment and arrest by the secret police. This was often followed by a disappearance, a euphemism for being sent to a labor camp. Lysenko's dismissal in 1955 brought a short reprieve which allowed for a burst of contact between East and West. Stalin's successor, Nikita Khruschev, had formally committed to the idea of peaceful coexistence. However, after the downing of an American U-2 spy plane Cold War tensions rose to a point where diplomats were no longer talking to one another.

Soviet scientists remained undeterred and continued their backchannel to promote cooperation with the West. The first big international scientific meeting—the Fifth Congress of the International Union of Biochemistry (IUB)—took place in Moscow in August 1961, a few months after Yuri Gagarin became the first man to orbit the Earth.[7] Western scientists realized they knew nothing about Soviet science and curiosity into scientific research behind the Iron Curtain piqued.

The Biochemistry Congress brought more than five thousand foreign biochemists to Moscow; including eleven hundred American biochemists accompanied by their spouses. Francis Crick and James Watson were the guests of honor following their discovery of the double helix. Their findings confirmed that DNA served as the carrier of genetic information.[8]

The Congress became memorable for a totally unexpected reason: the presentation by a young American scientist, Marshall Nirenberg, at the National Institutes of Health (NIH). At first he did not receive much attention. Nirenberg was not part of the "club" of eminent biochemists, and NIH was then thought of as an insignificant government institution. In a room with only a few people present, he spoke of his success in deciphering the genetic code and reading DNA. Crick and Watson had given voice to a theory that base-pairing of nucleic acids was a copying mechanism for genetic material, and that it provides the clue to perpetuating that information when DNA is duplicated to make two daughter cells.[9]

A few years later, they made the connection with the work of Jean Brachet, a Brussels University scientist who found that RNA played a role in the manufacturing of the cell's protein. But Crick and Watson had not answered the question as to how. How did the order of a series of nucleic acids, each attached to a sugar and a phosphate, in other words nucleotides, denote genetic information? It was these rules for reading DNA that Nirenberg had revealed. One listener understood the immense importance of this message and asked the organizers whether Nirenberg could repeat his reading, this time in the great hall to hundreds of scientists. The audience was electrified. There was much excitement, with many voices exclaiming "He did it! he did it!".

A young student in literature, Harold Varmus, happened to be in the city visiting Moscow. His friend had attended the conference and gave him an enthusiastic report at the end of the day on what it meant to decipher the genetic code:

> Just as a language based on written letters would be impossible to understand if we didn't know the lengths of words and their meanings, it was equally hard to understand what DNA said or how a cell interpreted it.

It was this language that Nirenberg had begun to clarify. He showed that a synthetic messenger RNA made of only uracils can direct the production of a protein. It was the first piece of the genetic code.[10] Varmus recalled that he was so taken by the excitement that "I began to understand that something of fundamental significance had occurred." A few months later Varmus shifted his field of study from literature to science. He was later awarded the Nobel Prize in Medicine in 1989 for his discovery of the oncogene that produces cancer.[11]

All participants were eager to return to their laboratories and see how they could build on Nirenberg's findings. Most people were oblivious to the drama that was occurring not far from Moscow: the Berlin Wall was being erected at breakneck speed four days after the conference began. East German authorities suddenly had enclosed the city in order to prevent further hemorrhaging of people. West Berlin had remained a loophole in the divide between East and West through which more than 2.5 million East Germans had fled to freedom, but now this exodus had come to an abrupt halt. West Berlin had become a beleaguered island within Eastern Germany that like other countries in Central and Eastern Europe was separated from the West with electric fences, minefields and armed guards. Churchill's use of the Iron Curtain metaphor in a 1946 speech was eerily prescient. The Soviets were indeed ruthless and would do anything to keep the countries occupied by the Red Army since the liberation after World War II firmly under their control.[12]

The scientists were not aware that Lysenko had been re-instated in early August 1961. It was a paradox in the history of science that the Rosetta stone for the DNA code was unveiled in Moscow, home to its fiercest opponent. Lysenko was back in control of the nationwide network of research institutes concerned with agricultural biology and would extend his benighted influence until Khrushchev was removed from power in 1964. Within the Eastern bloc, it was advised not to use the term "molecular biology" as it was dismissed as a kind of bourgeois fancy.[13] Lysenko's return signaled a revival of the more intransigent factions of the ruling communist party. It seemed as though the countries behind the Iron Curtain were going to revert to an autarkic isolation.

The Academy of Sciences in Prague

After World War II, Communism flourished across Europe. In Czechoslovakia, the communist party had become particularly popular. Czechs were angry over their betrayal by Western countries that had allowed their country to be dismembered in 1938 in order to appease Hitler and had forced them to endure humiliating subjugation as protectorates of Nazi Germany.[14] The communist coup of 1948 found plenty of people eager to place their faith elsewhere.

The siren call was irresistible for František Šorm who became a member of the Czechoslovak communist party.[15] Like many Czech students, he was drawn to leftist ideas during his university years in the 1930s, even though Czechoslovakia was prosperous at that time and figured among the ten most industrialized nations. He was especially taken by the support that science received from the government in the Soviet Union. He often lectured on the subject, praising the Soviet model once he became a professor of chemistry. He remained loyal to the party irrespective of its leader and didn't flinch after Antonín Novotny formally became first secretary of the communist party and continued his predecessor's rule of terror.

Abiding by Soviet instructions meant dissolving venerable institutions and learned societies in order to make room for the creation of a new Academy of Sciences in Prague. Following the Soviet model, the Academy would conduct research separate from universities; it also offered a home to several chemistry institutes underscoring the rich Czech tradition in this field.[16] František Šorm was asked to help organize the new body, and became Managing Director of the Academy in 1952. That same year, he created a new Institute that he placed under the umbrella of the Academy, consisting of a group of chemical enthusiasts from the Technical University. The Institute's name included the words "Organic Chemistry" and "Explosive Substances" as if to warn unsuspecting visitors that experiments could on occasion go awry. It was later renamed the Institute for Organic Chemistry and Biochemistry (IOCB).[17]

In 1962, Šorm became president of the Academy, the same year the Cuban missile crisis brought East and West to the brink of a thermonuclear war. The Missile Crisis, however, did not seem to affect Czechoslovakia. Instead the country continued to remove barriers and ease travel restrictions after contact with foreign countries had been reduced to a minimum during the 1950s. The arts, literature and filmmaking sprang to life.[18]

Šorm led a rather ambiguous life. To him, Communism did not mean the blind adoption of Soviet concepts. Instead, he tweaked them, all while cultivating contacts within the leadership and navigating the halls of power with ease. When Otto Wichterle, another prominent chemist and inventor of the soft contact lens, lost his university post during a purge in 1958, Šorm deployed his political clout to rescue him.[19] A new institute was set up within the Academy tailored to suit Wichterle's talents: the Institute of Macro Molecular Chemistry. The separation of research institutions from teaching universities had one great advantage. It meant the institutes of the Academy could be safeguarded against mindless rules imposed on universities. Life as a medical student at a university in Bratislava can be gleaned from the memoirs of a scientist who fled Czechoslovakia at that time:[20]

> A general atmosphere of fear and suspicion permeated life at the university. Teachers and students suspected of insufficient devotion to the Communist regime were investigated by specially appointed lustration committees. Universities had to abide by the ideology of Trofim Lysenko, the terms 'gene' and 'genetics' were practically banned. In addition to legitimate medical school courses students were also required to take courses on Marxism-Leninism, meaning primitive Communist propaganda.[21]

In Western Czechoslovakia, at Charles University in Prague, students remember a whiff of revolt. The head of the Department of Biology, Bohumil Sekla, uncompromising with the politics of the day, taught Mendelian laws to packed lecture theatres of medical students, who were excited to experience public expressions of dissidence at a time when controversial opinions could have cost a man several years in jail.

One of Sekla's brightest students, Milan Hašek, went on to lead a department of genetics in the Academy's Institute of Biology. As an active member of the communist party, he joined the Lysenkoist establishment and broke with his former mentor. Hašek's research project on immunological tolerance brought him international fame but he lost his credibility when he tried to espouse Lysenko's theories.[22] An exercise he thought would be beneficial to obtain official backing and substantial funding to expand his department into a full-fledged Institute of Genetics.[23]

When František Šorm became a member of the Central Committee of the communist party, the highest organ of the party, as well as a member of its Ideological Commission, the Academy could freely spread its wings internationally. Šorm encouraged his institutes to freely exchange contact and publications with scientists from around the world. He ordered that the Academy's communications would no longer be solely in Russian and Czech, but also in English. His mantra, that "the character of science is international and belongs to the whole of humanity," permeated the Academy.[24]

He noticed that his Western colleagues were often accepting each other's postdoctoral fellows and carrying out experiments in other people's laboratories.[25] Šorm made sure to extend similar invitations, all while building a network based on world-class scientists he knew from his student days like Vladimir Prelog and Sir Alexander Todd, both Nobel Prize laureates. Researchers from all over the world including those from the newly decolonized countries in Africa and Asia came to Prague. The Institute became a lively multicultural beehive. To people from abroad, it was obvious that the man at the top, František Šorm, set the tone for the open mindedness and creative energy bubbling up in the Institutes of the Prague Academy.[26]

In 1964, when Brezhnev replaced Khrushchev, it seemed safe for a group of seventy geneticists to travel to Brno in the eastern part of Czechoslovakia. They attended major celebrations engineered by Šorm's Academy together with the Soviet Academy of Sciences around the hundredth anniversary of Mendel's first communication in 1865. The Soviet delegation made a point to lay a wreath in the Augustinian monastery courtyard on Mendel's newly restored monument, which had been removed from its pedestal after the Communist's coup in 1948 and kept in some barn.[27] The following year, Milan Hašek signed a petition against a dogmatic article propagating the Lysenkoist doctrine in a leading Czech newspaper. He did not know he would later have to pay a heavy price. Those were still the happy days before the Soviet Invasion.[28]

One of Šorm's more famous visitors was Carl Djerassi, an American with Bulgarian and Viennese roots who felt perfectly at home in Prague. Djerassi recounts his invitation to give a lecture in the mid-Fifties just after the Budapest uprising. He remembered Šorm as a dutiful communist technocrat with a large Stalin picture hanging behind his desk, "his cunning eyes seemed to follow me whichever way I sat during our conversation."

By the early sixties, after Šorm had become President of the Academy, the picture had disappeared. Carl Djerassi, father of the first oral contraceptive, combined his position as a company Research Director of Syntex with a professorship at Wayne University in Detroit.[29] When Djerassi became the Head of Stanford's chemistry department in 1960, he intensified his contacts with the Academy in Prague and welcomed quite a few postdoctoral fellows from Šorm's stable into his laboratories. Like his friend Djerassi, Šorm became an active Pugwashite and hosted in 1964 a Pugwash conference in the Czech spa town, Karlovy Vary.[30] The town was a playground for the communist elite, but would also become a temporary venue for concerned scientists wanting to bridge the gap between East and West.

František Šorm used this forum to openly argue that a small country needed access to Western technology in order to participate in the scientific-technological revolution. It was a radical change for Czechoslovakia. Until then, it had been the Soviet Union's most loyal ally; but the Czechoslovak economy, that once ranked among the ten most industrialized countries before the war, had become a disaster. The only path to salvation was cooperation with the West. Despite its misgivings, the communist party establishment acquiesced once it realized that the country desperately needed to catch up economically.

Trade and exchanges with the West were also a boon for the Academy since its goal was to combine research with industrial applications. Despite patenting just about every research result, the Institute had only concluded license agreements with companies within the Socialist bloc. Šorm believed that agreements with the West would bring in hard currency and allow the institutes to equip themselves with valuable new devices. In 1965, Czechoslovakia made arguably the most successful transfer of technology from East to West when the Academy sold the license to produce Otto Wichterle's soft lens to an American company.[31] Ironically, the invention made by a "socialist scientist" sparked interest into hydrophilic materials in the West and created a new field of study.

Another "first" was the agreement Šorm and Djerassi signed. The father of the pill had become interested in birth control for insects. Djerassi's pill had been instrumental in unleashing the sexual revolution of the 1960s and now he wanted to control the insect population. His new company Zoecon, with Syntex as its largest shareholder, convinced a whole group of researchers in Prague to work on insect growth regulators.[32] It was the first collaboration agreement between the Institute and an American company. Another scientist who began his career at Syntex, John Martin, became the driving force behind a major agreement that consolidated a critical triangular relationship between IOCB, the Rega Institute and Gilead Sciences, an upstart American biotechnology company two decades later.[33]

Chapter III
Strange bedfellows: a Czech chemist and a Flemish virologist

Poets and prose writers are like amphoras waiting to be filled with wine or water. Scientists do not wait for their amphoras to be filled; they search for faucets producing spurts of liquid. Only the passage of time tells whether ambrosia or vinegar has issued. The search for the faucet is what counts.

— Carl Djerassi

Auspicious omens

Antonín Holý felt lucky. After his graduation from Charles University and military service, he was able to pursue his passion for chemistry, which he had nurtured since his youth. A book on chemistry for children, which he stumbled upon in his parents' attic, had ignited his fascination. His mother, who possessed a phenomenal memory for numbers, and his father, an artisan who made beautiful locks and other tools, stimulated his calling. They built a miniature laboratory for him in a corner of the garden shed. With a burning interest in his father's instruments and machinery, he developed uncanny skills during the 1940s.

Since his father was a craftsman and not associated with capitalist intellectuals, Antonín Holý was never branded as a bourgeois element. In high school he often substituted for his chemistry teacher who regularly abandoned the classroom. The teacher, as a fervent music lover, preferred to attend choir rehearsals instead.

Life in the family village near Prague was simple. His childhood friend, Ludmila, became the love of his life. She would have liked to become a biologist but the communists considered it unproductive and allowed only very few students into this field. She had no choice but to reorient her studies and focus on chemistry for the food industry instead. Antonín and Ludmila married, had two little daughters and formed a close-knit family. They shuttled between their modest home in the village near Prague and a *chalupa*, a country house where they grew vegetables to feed the household when shelves in the shops were depleted and distribution of goods in the communist economy was lacking.

His brilliant studies in organic chemistry at Charles University brought him to the attention of František Šorm, who snapped him up to work on a doctorate in the IOCB.[1] Chemistry was considered apolitical and its students were not as closely supervised as in other faculties. Organic chemistry, the chemistry of substances found in living matter, was so dear to his heart that it was quite a shock and disappointment to be assigned to an oligonucleotide chemistry group. Holý described his feelings in one of his essays, "My life with nucleic acids":

> When I joined the oligonucleotide chemistry group in IOCB I knew nothing about nucleic acids at all. I was not particularly fond of biochemistry lessons during the happy days at the faculty and there was nothing to improve this affinity during the days of my PhD study in synthetic organic chemistry.[2]

Things changed once he was hired as a full time employee and paired with an ingenious laboratory assistant. A statuesque woman, authoritative and gentle at the same time, taught him everything in biochemistry as well as the synthesis of the building blocks of heredity.[3]

His supervisor, Jiri Smrt, had a name with only consonants—*Smrt* literally meaning "death" in Czech. A pun was never far behind especially when he made one of his frequent visits to the lab of Sir Alexander Todd in Cambridge. Jiri's co-workers pronounced it *Tod*, also meaning "death" in German.[4] Holý absorbed every new technique or procedure his supervisor had picked up in Cambridge as if by osmosis.

The IOCB in Prague became in the early sixties one of the three or four renowned places in the world where nucleotide synthesis was cultivated.[5] This attracted many researchers at the Institute. Among them was Marc Van Montagu, a doctoral fellow from the Ghent University in Belgium. He was encouraged to go to Prague by his university's president, who did not hide his communist sympathies and had befriended many intellectuals and institutions behind the Iron Curtain.

Van Montagu would later become one of the first plant geneticists, a pioneer in engineering transgenic plants resistant to insects. He also happened to be the first foreign scientist Holý had ever met. They quickly became friends. During the three months spent in the IOCB in 1963, Van Montagu regaled him with stories of famous Belgian scientists. His favorite story was about Jean Brachet, considered by many as the father of RNA. He had found a method to show that this kind of nucleic acids is present in all cells whether they belong to plants, animals or human beings.[6] Brachet also indicated exactly where in the cell RNA is localized and pointed out its ability to transform genetic material into proteins.[7] Brachet, as so many academics in his time, defended communism as a reaction to the xenophobic and authoritarian governments that held Europe in its sway before the war.[8]

After the war, Brachet's communist friends forced him to endure their own form of harassment. He was required to write an article in support of Lysenko for the local communist newspaper. When he objected, he was then invited or rather summoned to Moscow to personally meet Lysenko. Brachet was so appalled by his absurd theories and the pressure he was put under to subscribe to them, that he abruptly quit the communist party upon his return to Belgium. Leaving politics aside, Jean Brachet and his work with nucleic acids inspired a whole group of bright scientists. Among them was Marc Van Montagu.[9]

Later, Tony Holý would write that their happy acquaintance was an auspicious omen that foreshadowed what would become his lifelong friendship and scientific partnership with another Belgian, Erik De Clercq. Holý's postdoctoral stay in 1964 in Göttingen, West Germany provided another lucky break that introduced him to the West. By the time he was invited to come to the Max Planck Institute, Czechoslovakia's former number one enemy, West Germany, was becoming its most significant partner in trade and in science cooperation.[10] The growing importance of international scientific exchanges was a troubling paradox for Czechoslovak hardline communist authorities. They disdained the way that West Germany would attract Czechoslovak scientists with money and research possibilities but they also realized that the knowledge gained in these exchanges was invaluable for their country's progress.

Holý was one of the beneficiaries. Aside from Russian and English, he spoke fluent German and thus was regularly invited to visit the Max Planck Institutes in Göttingen, an important site of pilgrimage for nucleic acid chemists in Europe. Holý became close to Fritz Eckstein, one of the brightest chemists in Germany at that time, who ten years later introduced him to his future science partner, Erik De Clercq. Holý's time in Göttingen was a productive phase in his career. His friendships brought him back to Germany almost every year, or whenever the Czechoslovak authorities allowed him to leave the country.

His friends in Göttingen gave him an endearing souvenir to take home: a handful of cactus seeds. The seeds grew into a most unusual cactus collection that he treasured with great care for the rest of his life. His German friends also helped him squirrel away a little money in a foreign savings account. His earnings in four months were about as much as he could earn during four years of work in Czechoslovakia, thus allowing him to buy a breath of freedom that came in the form of a small Italian car.

While Czechs and Slovaks were enjoying the loosening grip of the authorities and "socialism with a human face" was taking shape, Tony and his wife could travel abroad provided they left enough money in a communist-controlled bank account. They travelled with their little car to the Polar Circle in Norway and to the Low countries near the North Sea, the Netherlands and Belgium, where they visited his old friend Van Montagu. In Brussels, they were eager to visit the grounds where the World Fair of 1958 took place, a mythical place for Czechs of their generation. For the first time since the communist coup, the talents and creativity of Czech artists had been showcased in such a masterly way that Czechoslovakia won the prize for the most beautiful pavilion of the World Fair. Czech style and swanky furniture were admired by the whole world. Those were the happy days before August 1968.

A young prodigy's path to science

Did Erik De Clercq ever want to become a scientist? The thought probably never crossed his mind as a child. Erik was born during the first years of World War II and lived on his grandparents' farm until the war was over. He was raised in Hamme, a village bordering the river Scheldt, in Northern Belgium. Erik's parents owned one of the biggest plots of land in the middle of the village, which they transformed into a garden of wild flowers interlaced with vegetable patches and orchards. As an only child, he was doted on by both parents. His father worked as an employee in a fertilizer plant and regularly took young Erik to visit his lab[11]. They often went on Sundays when nobody was present so he could see "the chambers where sulfuric acid was produced." In Erik's mind, the whole lab belonged to his father; the smell of sulfuric acid and mysterious canisters discovered on those Sunday outings triggered his lifelong passion for chemistry.[12]

His mother had her own tailoring workshop on the ground floor of the house where she trained several assistants in the art of dressmaking. Young Erik loved to entertain them with his encyclopedic knowledge of world geography. They called him their "geography tutor." His mother, the dominant force in the family, decided very early on that her son should become a medical doctor and often prevented him from playing with other children. It was the time of the polio scares and isolation seemed the only way to avoid contagion, but in Erik's case staying indoors meant being locked up in his room to study. He lived in a world of his own, as he did not have many friends. For a while, his main interests were plants and animals. He first admired Carolus Linnaeus, the great Swedish botanist, but then shifted his curiosity to the animal kingdom and to lions in particular.

Academic discipline, inculcated by his mother, led Erik to achieve strong results in school. With his excellent grades in Latin and Greek, his teachers suggested he become a priest. Religion was far from his mind when he graduated from Dendermonde high school with the highest marks as a *primus perpetuus*. Art history, and the renaissance painter Rafael in particular, was much closer to his heart. He had no idea what university studies to pursue since nobody in his family had ever gone to university. One day, a longtime friend who had just started his studies in medicine dropped by their house and complained about the heavy load of chemistry lessons. As soon as he uttered these words, Erik's choice was sealed. He joined the medical faculty at the Catholic University of Leuven and it became his new home. There, he could pursue his passion in chemistry and at the same time fulfill his mother's dream of becoming a small town physician.

Leuven was a whole new universe, a city filled with students and tradition dating back to the 1400s when the university was founded by papal decree. Erik took to it like a duck to water, thoroughly enjoying his introduction to the microbial world in the university lab. Erik De Clercq, however, was not yet interested in virology; he was preoccupied with lectures on chemistry and assisting his fellow students. He became known as the "chemistry tutor," always eager to guide his fellow students stopping by his door with their questions. There was no textbook to accompany the lessons of their biochemistry professor, so Erik's notes were in high demand. Publishing his notes made him very popular and turned into a lucrative venture for the student in charge of the sales, who later bought a new car with the proceeds.

In the third year of the medical studies, only a handful of students could gain experience in a private laboratory. Professors would ask their brightest students during the oral exams to join their research projects. Erik received such an offer from Piet De Somer, a remarkably talented professor who was to become the president of the university. Inexplicably, Erik refused his offer. Yet in order to work in a prestigious lab, a student had to have solid introductions, which were hard to come by. One had to set in motion a whole network of relatives and friends. A cousin of Erik's father, who was a prominent school inspector, enlisted the help of her colleague to arrange an introduction with a professor from the Brussels University.

He suggested Erik, then 19 years old, join the laboratory of Christian de Duve, who had discovered lysosomes.[13] It implied studying cell fractionation, but this did not appeal to an innate stubborn young man. De Clercq had molecules in mind, not cells. After having snubbed De Somer, he declined to work with Christian de Duve, a future Nobel Prize winner. Instead, Erik ended up in 1962 in the obscure Laboratoire de Chimie Hormonologique.[14] He was not allowed to work with steroid hormones like cortisone and estrogens, which were his favorite, but was assigned catecholamines instead[15]. Alas, his duties consisted of diagnostic testing rather than hormonal protein chemistry. It didn't help that Erik disliked the spectrophotometer[16]. Clearly research in a laboratory seemed to be the wrong track for him. Seeing how he languished for more than two years in a place where no other language was spoken but French, a friend of the De Clercq family suggested he join the Rega Institute. As luck would have it, during another oral exam, De Somer repeated his offer to Erik to join his laboratory. This time Erik accepted. And so in 1964, two years before he graduated, he made his first acquaintance with the Rega Institute. Erik remained hesitant about his crucial decision.

When Erik De Clercq got acquainted with the Rega Institute, the whole team was still basking in the glory of its polio vaccine. It radiated the confidence and self-assurance of a task well done, of dealing with a disease in the eye of a media storm. Erik didn't know much about his new mentor, Piet De Somer, and was rather intrigued by his connection with the RIT pharma company. Grooming his new intern, De Somer didn't make any amends for praising the practical values of research. He ingrained into Erik the belief that "Research should be of help to mankind and if possible generate profits for the university. Research without a purpose makes no sense." In order to steer Erik into the field of vaccines, De Somer captivated him with a story about the prevention of many infectious diseases. The secret was to be found inside the human body and its own defenses. Erik was to become acquainted with the age-old battle between man and microbes.[17]

His first assignment occurred at the time when the rubella virus, the "monstermaker," that provoked birth defects in newborns, was sweeping around the world. After causing havoc in Europe, it spread to the United States in 1964, causing one of the worst rubella epidemics ever recorded and infecting twelve million Americans that year. Erik's task was to set up a fluorescence-activated antibody detection system. He would inject rabbits with an attenuated rubella virus of the Cendehill strain and then measure the antibodies concentration in their blood. When he reported he had found a very high antibodies count, he was greeted by an elated boss who foresaw this could be the basis of a new RIT vaccine. Piet De Somer was right; it became another success story. It was the world's first rubella vaccine to hit the market.

The work on the rubella virus also found its way into Erik De Clercq's master's thesis, an excellent opportunity for De Somer to persuade this young intern of the value of research and of starting a career in science. He told Erik that "physicians dealing with patients have to work like dogs."

De Somer referred to his own father, a small town physician, whose hard work had brought him to an early grave. "The physicians work benefits only the individuals they are seeing," De Somer said,

> Scientific research on the other hand, when practically oriented, can provide a treatment or a cure for a great number of people, maybe even for all of humanity! Infectious diseases have been the cause of more than half of the deaths in the beginning of the twentieth century. As scientists were able to devise more and more vaccines against childhood diseases, the rates of infectious diseases have gone down dramatically. The postwar generation is given a very different outlook on life. Isn't it a tremendous feeling to be part of this endeavor![18]

Erik had received a first taste of De Somer's persuasive charm, but remained hesitant about following a path into science. Even after his graduation as a medical doctor with maxima cum laude he kept his options open. His new position in the Rega Institute was contingent on the provision that he could jump ship at any time and start a career in internal medicine instead. Erik was very blunt with his mentor and told him that the two-year internship in the virology department had not been very rewarding since it didn't involve any chemistry. De Somer, now in his forties, was rather amused by so much rebelliousness. He coaxed Erik to formally join the Rega Institute in the summer of 1966 with the vague promise he could work on "the chemistry of viruses." After he had been "dragged" into research on viruses, Erik made sure he could report directly to the boss. He realized that Piet De Somer was an exceptionally charismatic figure who infused his laboratory with great vigor. It would become much more than a place just to work; it became a way of life.

Chapter IV
The sixties in Leuven and Prague

If Politics is the art of the possible, research is surely the art of the soluble. Both are immensely practical-minded affairs.

— Sir Peter Medawar

Antiviral penicillin

Vaccines to prevent attacks from viruses remained the prime focus of the Rega Institute throughout the sixties. Piet De Somer, however, wanted to look beyond prevention and search for a treatment. Once a viral disease had developed, there was no medicine available in those days.[1] Blocking a virus before it multiplies inside its host became De Somer's other fixation.[2] He took a keen interest in a young Scottish virologist, Alick Isaacs, who was working at the National Institute for Medical Research in North London.

Isaacs and Jean Lindenmann, a Swiss doctoral fellow, had discovered a new biological substance that could interfere with a virus replication. The two had been testing the behavior of chick cells after exposing them to a killed influenza virus. A substance appeared in the cells that prevented live influenza virus from growing. They had identified the substance as a protein but did not yet know whether it was produced by the virus or by the cell.

They named the protein "interferon" for its mysterious interfering activity.[3] Isaacs and Lindenmann explained in their articles, why no antibiotics to kill viruses had been discovered so far:

> To a large extent this is because viruses are extremely small parasites which are obliged to live inside cells, and it has not been possible to find a substance which would stop viruses from growing without at the same time harming the host cells. Interferon is the name which has

> been given to a new substance which prevents the growth of a number of viruses without apparently causing any gross damage to the cells. Interferon does not kill the viruses, but stops them from multiplying.[4]

Isaacs sent their articles to every prominent virologist he could think of. He received enormous press coverage in part because he had tagged the new substance with a powerful cultural symbol, the potential of becoming an "antiviral penicillin." The comparison reopened wounds of the penicillin trauma in the United Kingdom. Penicillin was still perceived as a British discovery that had been given free of charge to the United States during the war, while the British people on the other hand had to pay royalties to American commercial firms for every gram of penicillin they sold to the British market.[5] The British Medical Research Council did not want a repeat of the penicillin-affair to be repeated. The National Institute for Medical Research was not to waste any time and start a collaboration with three pharmaceutical companies working in the UK.

The international virology community was not as receptive. The cool reaction from eminent virologists was striking. Some started nicknaming interferon "misinterpreton" meaning the substance was probably a leftover virus particle, an abortive product of virus multiplication. Isaacs and Lindenmann, both physicians with only a modest knowledge of chemistry, were heavily criticized. Their vagueness as to the nature and mode of action of the novel substance made it extremely difficult to reproduce the results they had reported.

Isaacs's work nevertheless caught the attention of the recent Nobel Prize winner, John Enders.[6] He had also observed a protein blocking virus multiplication. At first he had called it an "inhibitory factor" but was now re-naming this biological fact as "interferon." This came as a tremendous boost for Isaacs. The event was witnessed by one of De Somer's co-workers, Edward De Maeyer, who had been sent as a postdoctoral fellow to Enders's laboratory in Boston.[7]

Now that interferon had been endorsed by such an eminent personality, it received an enthusiastic welcome in Leuven. Isaacs was hailed a hero and an honorary doctorate from the Catholic University was bestowed upon him in early 1962.[8] The ritual and festivities around this award underscored the new direction the Rega Institute was to follow. The Institute was being retooled to become one of the premier interferon centers on the continent. The buzzwords "antiviral penicillin," and "broad spectrum magic bullet" against many viral diseases, were music to the ears of De Somer. Would he be able to repeat his earlier success when, unburdened by royalties, RIT was able to produce its own penicillin?

This what was probably on his mind when he travelled to Smolenice Castle near Bratislava, Slovakia in 1964 to attend the first international conference on interferon. A young scientist, Jan Vilček, who had gained quite some traction in the interferon community after the publication of his paper in *Nature* in July 1960 had taken the initiative. Interferon was still a very small field at that time and probably half of the world's active researchers had travelled to Smolenice. Among the forty-odd participants were prominent interferon researchers from the United States, France, Finland, many participants from Czechoslovakia, and the Soviet Union. De Somer represented Belgium, he was accompanied by Edward De Maeyer and his wife, whom Jan Vilček had befriended on one of his rare trips outside the country.

Vilček recounts the Interferon Conference in his autobiography and mentions a shocking moment that in his mind marked the conceptual birth of commercial biotechnology. It was De Somer's relaxed attitude towards monetizing research that caused consternation:

> One evening, conference participants gathered to play a social game. A moment that has stayed with me in particular was when Piet De Somer answered the question what would you do if you were to discover an effective cure for virus infections. I would sell it, he said without hesitation.

He also recounts all the details about him plotting his defection with his Belgian friends without being noticed by the omnipresent secret police. A few weeks after the interferon symposium, the communist authorities surprisingly granted Vilček and his wife permission to travel for a weekend to Vienna. They never went back to communist Czechoslovakia and wandered through Belgium and Germany before arriving in the US in 1965 as penniless refugees. Jan Vilček was welcomed at the NYU School of Medicine and became a prominent American scientist. He went on to spearhead some key advances in the research of interferon and later into the tumor necrosis factor that led to new treatments for a wide range of autoimmune and inflammatory diseases.[9]

Systematic initiatives to produce interferon from human cells were only started around 1966. De Somer and his group opted for interferon in human fibroblasts, cells usually derived from abortion tissues or from the human foreskins obtained after circumcisions, a routine procedure for baby boys in the United States. Growing fibroblasts cells in the laboratory was no mean feat; it proved both laborious and expensive. Kari Cantell and his group in the Finnish Red Cross Blood Transfusion Service made interferon from human white cells or leukocyte interferon.

Neither De Somer's group nor Cantell's work received much attention at that time because the quantities produced were still not enough to test interferon on humans.[10] Interferon's credibility and the hopes to develop a useful drug sank to a low point.[11] Many scientists began to wonder whether there was an alternative. They recalled that Isaacs himself had suggested the possibility that a chemical compound could stimulate the human body into producing its own interferon, thus circumventing the laborious process of having to make interferon and then inject it. In the US, many had shifted from the administration of interferon or exogenous interferon to the induction of interferon, the endogenous interferon.

Piet De Somer wanted to pursue both avenues, continue his experiments with production of interferon while trying the induction approach at the same time. Two of his senior assistants were on extended leave in the US and had left a whole group of underemployed lab workers behind. It was an obvious choice to dispatch his new doctoral fellow, Erik De Clercq, to lead these eager technicians in exploring the induction of interferon.

De Clercq first wanted to prove interferon existed. After injecting rabbits with sindbis, a virus harmless to humans, he noticed interferon in their urine. This observation was embraced with great enthusiasm by De Somer. It showed that interferon was not something esoteric, its existence was proven by its low molecular weight passing over the rabbit's kidney threshold. It became the topic of Erik's first publication. Although, as was customary at that time, De Somer received most of the credit as the head of the Institute.[12]

De Clercq then began to scrutinize the products which medicinal chemists had sitting on their shelves to see whether any of them could stimulate interferon. Very early on, he found some polymers, or clusters of atoms, that could induce interferon in the human body. It was another eureka moment that provided yet another topic for a new publication and his doctoral thesis as well.[13] De Somer presented Erik's findings at a congress in Fort Lauderdale, Florida, in late 1966 where he renewed his acquaintance with the rising stars of American Interferon research.[14]

Tom Merigan in Stanford had been on the same wavelength as De Somer since both were among the first to make interferon from human fibroblasts and now both labs were exploring synthetic inducers of interferon. Merigan had already shown that interferon could be induced by synthetic polymers.

Piet De Somer also took an instant liking to Maurice Hilleman, the towering and imposing research director of Merck, the pharmaceutical giant. De Somer was in awe of Hilleman and shared the same mischievous sense of humor. They became close friends and Hilleman eventually profoundly influenced research into interferon.[15]

Spring turns into winter

The democratization process fueled growing criticism of the country's economic stagnation that had been brewing all through the sixties. The best Czechoslovak movies reflecting social and political tensions appeared in that period. Some of the films are satirical tragicomedies displaying the dry wit so typical of Czechoslovakia.

When another politician, Alexander Dubček, rose to power in the first months of 1968 to replace Novotny, the Prague Spring was reaching a climax. Dubček attempted to reform the regime. The relaxation of censorship brought the whole country into a heady, happy atmosphere but became a lightning rod for Soviet wrath over so much intellectual freedom. "Socialism with a human face" also infected the Academy of Sciences. Compared to social scientists that were caught up in the revolutionary fervor of the Prague Spring, natural scientists preferred to remain apolitical.

Otto Wichterle, the head of the macromolecular Institute, clearly became the exception to the rule. He was leading a group of dissenters who demanded recognition of a trade union of scientific personnel to insure internal democracy and to neutralize bureaucratic interference with their research. A core of historians and sociologists, with the notable input of Otto Wichterle, initiated a famous manifesto. It called on public denunciation of the secret police spies and for citizens to support the Dubček government, with arms if need be. It was published by an activist writer and noted rabble-rouser in a literary journal in June of 1968 and became known as the *Two Thousand Words Manifesto*. With its call to arms, it was the straw that broke the camel's back. This time, it provoked the Soviet Union into action. On the night of August 21, Soviet tanks and forces from other Warsaw Pact states rumbled across the cobblestones and invaded the Prague capital.[16]

The presidium of the Academy responded the next morning with a radio broadcast condemning the invasion.[17] In the first months of the Soviet occupation, there was still hope that the Dubček government would survive. František Šorm kept pushing reforms to gain greater autonomy for the Academy. As a member of parliament, he voted with the majority of deputies to condemn the August invasion as an unwarranted and hostile act and refused to vote for the right of the Soviet troops to stay on Czechoslovak soil.

In April 1969, when Alexander Dubček was replaced by the notorious hard-liner Gustav Husàk, retribution against the Academy began in earnest. Šorm sent a letter asking for support to his colleagues in the Academies of Science of all the Warsaw Pact countries. It was to no avail. He was dismissed along with ten other members of the Council of the Academy. The next president, a specialist in automation technology, soon issued a decree that would impose his short-sighted vision on the Academy:

> The socialist scholar does not waste time or means on developing theories which are not socially useful, nor on writing works which solve nothing and do not occupy a place in the list of social requirements.[18]

Next, he put in place the emergency measures against dissenting scientists. By June 1970, all Directors of the Academy's 138 institutes were dismissed and replaced by opportunists and party hacks. Outstanding scientists like Šorm and Wichterle were placed on a renewable contract for employment, ranging from a few months to three years depending on an assessment by the party. They were prohibited from publishing, travelling and reading foreign books or journals.

The previously liberal atmosphere quickly evaporated and was replaced with sullen management, countless restrictions and exasperating regulations. The one-sided orientation toward Soviet science that had prevailed in the 1950s was back. The wrath of the government against a scientific institution that harbored reformers and dissidents was vicious. Eventually, the Academy was placed under the control of the Government and the Party. The budgets of the Institute were drastically reduced, foreign travel and participation in scientific congresses became severely restricted.

For Antonín Holý, the humiliation of his mentor came as a grave shock. He vowed he would do everything to restore the honor of the man he so admired. His first priority was to preserve Šorm's creation, the IOCB.[19] When equipment of the laboratories became painfully scarce, Holý was most creative. All the skills he had learned in his father's workshop were called upon to blow glass canisters and to weld his own tools. He was no longer allowed to make trips abroad with his wife and children, but he would bring home extensive slide shows from each of his foreign visits. The whole family would share and analyze his observations.

Holý was not to be subdued. Encouraged by the recognition he had gained abroad through his publications, he concentrated on his scientific work. The making of phosphonates was becoming his trade mark.[20] He was aware that the leaders of the Academy had been admonished by the government and the communist party to make the institution economically productive and that chemistry was considered as something that could be useful for industry. Just about every one of his achievements, like those of other practitioners of the natural sciences, was placed under patent in the hope that they could be sold abroad to gain badly-needed hard currency.[21]

The Institute of Genetics, on the other hand, was viewed by the new leadership as totally unproductive. Hašek had been expelled from the Party and demoted from all his functions in the Academy. The famous immunogenetic school he had built was reduced to a shadow of its former self. More than twenty-four of his young scientists fled and dispersed to various countries around the world where they were to make a considerable impact in the field of immunology.

The return to dogmatic communism and its gray strictures was presented by the authorities as "normalization." Tens of thousands of people emigrated, leaving family and once-flourishing companies behind. People learned to keep their mouths shut in front of some communist zealots and those who were suspected of being secret police informers. After 1969, the Party blacklisted most of the country's best movie Directors, but simultaneously wanted film production to continue and be perceived as successful. Living in such contradictions sapped the morale of a nation during the seventies, and Czechoslovakia entered a period of paralysis.

In other parts of the world, the upheavals of 1968 meant something very different from the *Prague Spring*. Student protesters in Paris were calling for leftist governing regimes that if put in practice would be even more oppressive than those of Eastern Europe.

In Berlin, student riots broke out over the killing of a leftist student leader, Rudi Dutschke.

In Leuven, the student rebellion coincided with growing linguistic tensions in the centuries-old bilingual university. It also had an element of reaction against organized religion. The Catholic Church's clergy and hierarchy exerted a tight grip on Flemish villages where many students and professors of Leuven had originated.

In the United States, in 1968 Martin Luther King and Robert Kennedy were both assassinated. Riots erupted in more than a hundred cities and merged with protests against the Vietnam war.

Chapter V
Enzymes: the secret of life as chemistry

Nothing is sacred in science; you give up the old when you find something new that is better.

— Thomas Rivers

At Stanford, a new world opens up

It seemed the whole village of Hamme had come to the airport to wave the newly-wedded couple goodbye as they embarked on their trip to San Francisco. Friends and family were very proud of Erik De Clercq, the bright young medical doctor, but also of his lovely bride, Lili, who had grown up in the village apothecary just around the corner from Erik's family home. Lili herself was an accomplished pharmacist. It was a marriage made in heaven, bathed in an aura of chemistry. The honeymoon kicked off to a thrilling start. Neither Erik nor Lili were experienced travelers and flying in a plane was an absolute novelty. The last leg of the trip, from San Francisco airport to Palo Alto by helicopter, filled their stomach with butterflies. Gazing down at the verdant landscape, they were totally overwhelmed by the beauty of the hills undulating like the Tuscan countryside in Italian renaissance paintings.

De Clercq had received a fellowship to come to the Stanford Medical Center in 1968 and work in the laboratory of the Infectious Diseases division.[1] The Director was the quick-witted Tom Merigan, one of the youngest professors at Stanford at that time and only a few years older than his postdoctoral fellow. Merigan and his wife could not have been more welcoming as they helped the honeymooners settle into their new home in the California hills. The balmy weather, the sun burning off the fog that rolled in from the Pacific Ocean, a house surrounded by exotic flowers and a swimming pool made it seem like paradise.

Stanford University was more sedate than Berkeley, which just across the Oakland Bridge had become the epicenter of student protests against the Vietnam War. The flower power movement and hippies in nearby San Francisco seemed far away. Erik De Clercq was consumed by his research and the task of absorbing new knowledge. His bride was almost as passionate as he was, transcribing during the night with an old fashioned typewriter the notes he had made during the day.

Merigan and De Clercq were fascinated by the startling news that had just come out of the Merck laboratory.[2] Maurice Hilleman had found that certain nucleic acids could induce interferon both in cell cultures and in laboratory animals on a much larger scale than was ever thought possible.[3] This had an immediate effect on just about every laboratory that was investigating viral diseases. A decade after the discovery of interferon, most scientists were eager to work on the *induction* of interferon, which became the new wave in research.[4] The ultimate goal was to find clusters of nucleic acids or polynucleotides that could be used as drugs in humans.

Merigan had abandoned his *production* of fibroblast interferon in order to jump on the inducer bandwagon. Chemists in his wide network of contacts sent him all kinds of compounds to be tested in his lab. De Clercq seized the case like a detective on a hunt. During one of his lab tests, he found that the compounds made by Fritz Eckstein at the Max Planck Institute in Göttingen were surprisingly powerful inducers.[5] The discovery was published by De Clercq, Eckstein and Merigan in *Science* and in the *Chemical and Engineering Magazine* which won them instant acclaim.[6]

A patent for these interferon inducers was not far behind and preparations were made for the legal paperwork to be signed together with Fritz Eckstein. In the legal labyrinth at Stanford, Erik's name as co-inventor on the patent had been left out and was only reinstated after Eckstein's insistence. But the compound never became a drug. It never went further than a hot and uncomfortably humid visit to Philadelphia to meet with Wyeth, the company which bought the license but never produced the drug. The frustrating effort had nevertheless yielded one blessing: Erik had acquired a new friend, Fritz Eckstein, who introduced him a few years later to chemists who would open new doors in his professional life.

In Tom Merigan's lab, De Clercq not only learned to protect his patent rights but also acquired the all-important habit among scientists to "publish or perish." For Tom, each and every one of De Clercq's findings was worthy of publication. If an editor was slow in responding, Merigan would urge Erik to reach out to a more prestigious journal. Not surprisingly, two years at Stanford yielded twenty-five of De Clercq's publications, each of them appearing in distinguished scientific journals.

Stanford harbored many other delights. Thanks to his fanatical work schedule, working at night and on weekends, Erik earned unfettered access to the scintillation counters in the laboratory. In his free time, he would attend lectures given by some of the giants in molecular biology, almost on a daily basis. Arthur Kornberg's lectures left a deep impression.[7]

Borrowing from his experience as a ship doctor and a nutritionist at NIH, Kornberg sprinkled his lectures with philosophical insights but always circled back to enzymes. The yeast cell, which is responsible for the birth of modern biochemistry, can convert sugar or starch into wine or beer. The ways it gives champagne its sparkle and bread its leavening effect have led biochemists toward a deeper understanding of the molecular basis of cellular behavior.

Kornberg taught his students that enzyme is Greek for yeast,

> You have to know the actors in order to understand the plot. And the actors are the enzymes. They are the mini-chemists, the devices by which a biological phenomenon takes place, whether it is the legendary question of alcohol fermentation or how a firefly comes to luminesce. In naming each of the thousand and more enzymes that have been discovered, the suffix '-ase' has usually been added to the chemical process it catalyzes.[8]

By studying enzymes, Kornberg unraveled the complex chemistry of DNA replication:

> DNA is simply the construction manual that directs the assembly of the cell's proteins. It also serves as a template for replication in order that the DNA of two daughter cells will be identical to that of the parent.

He was awarded a Nobel Prize in 1959 for his discovery of DNA polymerase, the enzyme that makes DNA. He was quoted as saying that "DNA itself is lifeless; what gives the cell its life and personality are enzymes." Later, he synthesized DNA which possessed the genetic activity that created "life in a test tube," as coined by the press, and opened the way for his postdoctoral fellow, Paul Berg, to develop *recombinant DNA*.

Kornberg created a congenial atmosphere both in his laboratory and in the lecture hall. He was kind, witty, and adored by all his students, some of whom he launched into illustrious careers. With his very didactic, almost poetic skills, he proved that teaching and exquisite research can go hand in hand and even reinforce one another. He had a decisive influence on Erik De Clercq's understanding of biochemistry and his love of teaching.[9]

Tom Merigan offered Erik the possibility to stay at Stanford and become the head of a new clinical virology department. It was an offer he would have loved to accept. Piet De Somer, however, visited the United States regularly and kept a close watch on his postdoctoral fellows, like the Argus giant with one hundred eyes. He was all too aware of the very attractive conditions they were working in, particularly those in Stanford. Erik De Clercq was reminded of his contract; the time at Stanford was up. He recounted it as a Faustian pact with the devil. He was left with no choice but to come back to Belgium.

Erik and Lili left Stanford with a heavy heart. They were allowed to delay their return as De Clercq had been invited to give lectures on the East Coast in the DuPont experimental station in Wilmington, Delaware, at Johns Hopkins in Baltimore, Maryland and at the NIH near Washington, DC. He made many new friends, some of whom became his future co-workers.

The Rega Institute cuts corporate ties and goes its own way

Back in Belgium, Erik found that things had changed rather drastically. The search for a worldwide distribution of RIT's rubella vaccine had attracted the attention of a major pharma company, Smith Kline and French.[10] The small and dynamic RIT was now becoming part of the SKF empire.[11] The new pharma giant was not interested in the Rega Institute's research into interferon and subsequently cut ties with the Rega Institute.

On a political level, the fracturing of the six hundred year-old Catholic University in Leuven left a deep scar. It unleashed a nationalist fervor for an independent Flanders that began to mount as the Flemish community grew wealthier and stronger. The by-product was a diminished role for the Catholic Church in the management of the university. It reinforced the University's historical independence of scientific thought. One of its proudest achievements—paradoxical for a Catholic university—was the development of the "Big Bang theory" in the 1930s by the Leuven priest Georges Lemaître.

Piet De Somer was entrusted with a new role as the president of the Flemish university. As the quintessential figurehead representing both the old tradition and the new wave of secularization, he was at the peak of his fame. The charisma, the vitality, the sheer energy that radiated from him electrified any room he entered. In the Rega Institute, his presence was felt even when he was not physically there.

But somehow, things were not the same. Several French speaking researchers had been hired away by RIT/SKF or moved on to other universities. Erik was assigned to an empty lab left with nothing but broken glass and worn furniture. It was hard to readjust to the cool Belgian climate and the competitive atmosphere. Above all, he sorely missed the informal Californian work environment. To make things even worse, the military draft and life in the barracks was awaiting him.

Only De Somer's contagious optimism and the companionship of a devoted technical aide could dispel his misery.[12] His other consolation was his favorite magazine, the *Journal of the American Chemical Society*, and his correspondence with some of its authors.

One of these letters was addressed to Dr. Bernhard Witkop, the Director of the Chemistry Department at the NIH and one of the more authoritative members of the American National Academy of Sciences. Was it baffling naiveté or plain guts to ask a famous scientist for access to his compounds? Undaunted, Erik asked him to send his polynucleotide products to the Rega Institute so they could be tested for interferon. First he received a polite no, but then, several months later, out of the blue, Witkop sent him a request to test several compounds within the short time span of two weeks. It was the beginning of an intense collaboration.

In 1971, production of interferon had gained the upper hand and induction with polynucleotides was moved to the backburner. Piet De Somer organized an international conference on interferon. Tom Merigan was one of the star participants. He had come to realize that inducers were, after all, not very promising.[13] In the margins of the conference, he developed a plan to start clinical trials in Stanford for cancer patients with shingles. This viral disease, painful but innocuous for healthy people, was a calamity to immune-depressed patients. Merigan preferred the Finnish leukocyte interferon, which was the only source available in enough quantities to serve a few patients.[14] The Rega Institute still had all of its hopes invested in the fibroblast production, which it claimed was much purer than the Finnish white blood cell interferon and did not carry the risks of all kinds of infections.

De Clercq did not play a significant role in the conference, he was distracted with teaching biochemistry in the medical faculty and his new research topics.[15] He was fixated by a new enzyme discovered by two American scientists, David Baltimore and Howard Temin, who had been working independently from each other but had arrived at the same conclusion. Even more remarkable, their two papers were published in the very same issue of *Nature* in June 1970. They had found that certain RNA viruses, often acting as a silent passenger, but sometimes causing leukemia or solid tumors in animals, had the help of a unique enzyme, reverse transcriptase. It was an enzyme that only very specific RNA viruses carried.

Not until the beginning of the 1950s were scientists able to distinguish RNA viruses from DNA viruses. Now there had been an enzyme discovered in a certain kind of RNA viruses that made them behave in an unusual way. These viruses first transcribed their genetic material into DNA and only later back into RNA. An infection by this kind of RNA virus, later renamed retrovirus, had the same effect as an infection by a DNA virus, and thus could stay in the organism indefinitely. This activity was called reverse transcription. It totally upset the prevailing dogma. In those days, it was believed that DNA *always* transcribed into RNA in order to produce protein in the cell. The newly discovered enzyme reversed that order. It directed the information flow from RNA back to DNA, or from a gene's message back to a gene. With reverse transcriptase, every RNA molecule in a cell could be used as a template to build a corresponding DNA.

In the Proceedings of the National Academy of Sciences (PNAS) it was speculated that this virus was at the origin of all cancers.[16] Erik was totally absorbed by this tantalizing hypothesis. If the cause of cancer was indeed a virus, a certain kind of RNA virus, then surely a way could be found to contain the virus and thus contain cancer. Discreetly, and without the help of any technician, he repeated all the biochemical experiments described in the *Nature* papers. He then validated these results with real life testing on mice. Building on the experience he had gained in Stanford, he used mice to grow a virus causing leukemia, then peel off the cancers and re-inject them into other mice to see if they would in turn grow cancers within ten days.

He was overjoyed that his "test tube" results confirmed there was indeed such an enzyme behaving as reverse transcriptase. This ritual would be repeated over and over, every time with different reagents to see whether any compound could inhibit the cancer cells. One day, he found a substance that is used against sleeping sickness in Africa, Suramin. It was a most curious experiment. It destroyed the virus in the test tube but had no effect on the cancer cells. His dream of finding a cure for cancer was instantly shattered, at least for now.

Chapter VI
From interferon to nucleosides

The progress of science is strewn, like an ancient desert trail, with the bleached skeletons of discarded theories which once seemed to possess eternal life.

— Arthur Koestler

A first encounter with nucleosides

The dwindling interest in the induction of interferon suited Erik De Clercq very well. De Somer allowed him to broaden his interests and travel. Erik eagerly jumped on the occasion to go to Bulgaria in 1971 to attend an annual conference of the Federation of European Biochemical Societies (FEBS).

It was Erik De Clercq's first encounter with a relatively new organization at that time, FEBS, the brainchild of British biochemical societies. Most of the members of those societies had taken part in the famous International Union of Biochemistry (IUB) congress held in Moscow in 1961. They still reveled in the fact that the congress had brought them into personal contact with so many of their counterparts. The problem was that the IUB congresses were held only every three years, with the next conferences to be held in New York and then in Tokyo. Such long-distance travel was not easy for younger biochemists who found it hard to keep up with foreign colleagues. So the idea of organizing a Federation of European Biochemical Societies (FEBS) began to germinate within the Oxford and Cambridge societies. They conceived a platform in 1964 with annual congresses, alternating between countries in the East and West. An important principle was that political, national and territorial considerations would be ignored.

De Clercq was invited to Varna, a seaside resort on the Bulgarian Black Sea Coast, where the 1971 FEBS conference was held. Advertisements had promised a "Communist Riviera" with golden beaches, so he took his wife Lili along. Reality was rather different. The grayness and dirt matched the images of gloom and doom that filtered through other stories about Eastern Europe.

It was nevertheless in Varna that De Clercq met one of the most prestigious American chemists, Bill Prusoff, who was based at Yale University. Prusoff had acquired quite some fame for the nucleosides drug he had synthesized in 1959. It was supposed to be a cancer fighting agent, but a biologist, Ernest C. Hermann, later discovered it was in fact an antiviral. But it was the chemist who synthesized the drug, not the biologist, who would be hailed as the *father* of the first antiviral drug, a nucleoside. And thus, Prusoff entered medical history.

It was also in Varna that Erik De Clercq met David Shugar, the head of the Biophysics Department at the University of Warsaw. Shugar was an important player in the organization to bring scientists from East and West together. He had organized the first FEBS meeting. The meeting in Warsaw in 1964 was attended by more than a thousand scientists. It was considered nothing short of a miracle to bring so many distinguished scientists together in a country behind the Iron Curtain.

David Shugar had a warm heart for anything Belgian, perhaps due to the fact that Ghent University had awarded him an honorary doctorate.[1] The pipe-smoking Canadian had a tumultuous past. He had been charged in the early 1950s in connection with espionage activities. The affair was triggered by the defection of a cipher clerk in the Soviet embassy in Ottawa who accused Shugar of overly close contacts with Soviet diplomats. Even though he was never convicted of any wrongdoing, the investigations and the publicity damaged his career irrevocably. He was tarnished with the spy label despite being guilty of nothing more than "infatuation with communism."

After he moved to France, authorities there started to harass him, so he fled to Belgium. In Brussels, the wealthy Errera family and their legendary salon gatherings introduced him to the school of Jean Brachet and his nucleic acid chemistry.[2] Brachet influenced Shugar profoundly. Still, he did not feel safe in Belgium and the local police soon started questioning him as well. McCarthyism had long tentacles and only his native Poland could grant him a safe harbor. When he was offered a position at Warsaw

University, he gladly accepted and became perhaps the only scientist during the Cold War who fled from West to East.

Erik De Clercq and David Shugar became close friends. They immediately started to work out a plan whereby Shugar would send him polynucleotides to be tested and analyzed at the Rega Institute. Shugar, who was a physicist by training and a recent convert to biology, found the chemistry side of producing polynucleotides rather tedious and cumbersome. He suggested sticking to a simpler method of manufacturing. Why not synthesize nucleosides and put them to a test for interferon? De Clercq agreed reluctantly. He did not realize then how nucleosides were going to fundamentally change his future work.

A fateful meeting in Göttingen

Bernhard Witkop, Head of the Chemistry Department at NIH, had started a series of collaborations with Erik De Clercq, but liked to delegate many tasks. His deputy, Paul Torrence, regularly came to visit Erik in Leuven to see how things were progressing. This time, Torrence needed to stand in for his boss who was too busy preparing for a lecture in Japan. Witkop was meticulous to the extreme; he liked to read his text in classical Japanese.

Torrence arrived on the first day of May 1976. Leuven was primed for a festive May Day parade and the many red flags certainly made his visit to the Rega Institute very colorful. He wanted to rehearse the presentations he was preparing to deliver at a symposium in Göttingen that he would attend with De Clercq. They drove his old car at a leisurely pace so Torrence could enjoy the German landscape. Their destination was tucked between the Harz Mountains and the Weser River somewhere in the middle of West Germany. Seeing the centuries-old timbered houses behind the old town walls of Göttingen was quite a thrill for both. The little town, home to a famous university and several Max Planck Institutes, had fostered forty-six Nobel Prize winners.

Two of the Max Planck Institutes had selected about thirty chemistry researchers to spend a few days in Göttingen.[3] The small symposium on *Synthetic Nucleosides, Nucleotides and Polynucleotides* attracted some of the finest nucleic acid chemists from both the US and Europe, with only one medical doctor present, Erik De Clercq.[4] His friend Fritz Eckstein, with whom he shared a patent and several publications, was the likely instigator behind his participation.

De Clercq felt more at ease after spotting another friend, the perpetually smiling and pipe-smoking, David Shugar. They had established an intensive working relationship ever since they first met in Varna and had already published some twenty papers together. Shugar sent bright Polish researchers to the Rega Institute; it didn't hurt that they were female. They had given lectures at each other's universities. When De Clercq flew to Warsaw, it was immediately obvious that Shugar was in good standing with the communist authorities. In the airport, he was able to whisk his visitors through customs and immigration in no time. Despite his powerful connections, Shugar lived in surprisingly modest surroundings; in a small flat filled with books, not far from his lab.

David Shugar was well known in laboratories in both the East and West. With his Canadian passport and Polish residency card, he could easily navigate through the Iron Curtain. Thanks to his Jewish roots, he was also a welcome guest at the Weizmann Institute in Israel. The Israelis provided him with chemical compounds, the first nucleosides, that otherwise would be unavailable in Warsaw. Shugar developed a special bond with the head of the Weizmann Institute, Ephraim Katzir (Katchalski), a biophysicist who became the fourth president of Israel in 1973. Katzir also was his protector at times when the political climate in Poland turned anti-semitic. Shugar was a cross pollinator; he had quietly guided Erik's curiosity into the world of nucleosides. Both were now tackling the question of whether nucleosides could possess antiviral activity.

In Göttingen, every participant gave a presentation on their recent research. De Clercq's talk shed new light on fighting viral disease. Two well-known American scientists, John Montgomery from the University of Alabama and John Moffatt from the Syntex pharma company, were immediately interested in collaborating with the Rega Institute. With the exception of these few enthusiasts, most other chemists knew very little about antiviral drugs. Most of them believed that vaccines were sufficient to prevent virus infections and that there was no need for treatment.[5] They likened all viral diseases to a common cold: "if treated vigorously, it will go away within seven days, whereas if left alone it will disappear over the course of a week." Their knowledge did not extend much beyond using Bayer's aspirin to treat the flu and quinine for fever suppression.

So Erik De Clercq's presentation linking certain nucleosides to antiviral activity was met with a mixture of awe and incredulity. One participant, a crude Russian from Novosibirsk, loudly protested. It only heightened the interest of the soft-spoken Czech scientist in the room, Antonín Holý. Years later, he wrote:

> Erik has by far the best chemical mind among the M.D.s I ever met; he correctly estimated the potential of nucleosides as antimetabolites acting against cellular parasites. This encounter developed into a friendship which has had a decisive role in my professional life. [...] In those days I had absolutely no knowledge of viruses, their life cycle and pathological manifestations and I presume that many of my contemporaries must have felt the same. After all, this knowledge was at that time rather scarce anyway. The pharmaceutical industry paid it but a formal interest [...].[6]

The Max Planck Institute was where Antonín Holý had spent the only postdoctoral stay in his life. Through his short annual trips to Göttingen, his home away from home, he had kept close contact with his friends, especially Fritz Eckstein. Everybody knew him as "Tony." This regular commute between Prague and Göttingen had been possible throughout the sixties until the *normalization* of 1970 imposed a travel ban in Czechoslovakia. It was only after a major East-West conference in 1975, the International Conference in Helsinki, that communist authorities began to relax the rules again.

Over the years, Holý learned to mimic the German accent so well, it was hard to believe he was a foreigner. His famous ear for classical music must have aided his linguistic skills. His favorite pastime on his trips to Germany was visiting hardware stores. His toolbox was his pride and joy. He was always curious to explore the latest in German tools, everything from a screwdriver to more sophisticated instruments that would be handy to pack in a suitcase.

Holý's topic, the chemical and biochemical aspects of L-nucleosides, remained anchored in De Clercq's memory. Many years later, these analogues became anti-HIV drugs. Holý did not speak about the political hardship in Prague, but instead he described the limitations imposed on his work at the IOCB.

Due to the scarcity of starting materials, he had to milk an African snake to get fresh snake venom and extract enzymes commonly used for analytical chemistry of nucleotides. Holý and his colleagues had also been toying with the idea of growing Japanese carnivorous plants whose digestive juices were reported to contain precious enzymes. Somehow, they had been overheard by helpful diplomats at the Czechoslovak embassy in Tokyo. Later, embassy packages with enzymes would mysteriously arrive at the Institute.

De Clercq was very touched by the quiet and dignified manner in which Holý coped with adversity. He promised that as soon as he was back in Leuven, his technician would send all kinds of reagents to Holý's lab to replace the expensive imports the Institute in Prague relied upon. In turn, Erik would ask whether his new friend, Tony Holý, could send some compounds to Leuven to be tested for antiviral activity. This became a standard request that De Clercq would ask of every chemist he would meet. These gifts greatly contributed to the Rega Institute's impressive collection of compounds.

De Clercq had also caught the attention of a British scientist from the University of Birmingham in the UK, Richard Walker. He kept quiet most of the time but had his eye on De Clercq as a possible co-organizer of NATO's Advanced Study Institutes. The institutes, in turn, would lead the way for some groundbreaking collaborations among the participating scientists.

Bringing compounds to life

Holý and De Clercq did not waste any time after their meeting in Göttingen. A very courteous correspondence began immediately once they were back in their labs. The first compounds from Prague to be tested for antiviral activity arrived in Leuven a few weeks later.

They had to be mailed through a special clearance company that had been entrusted by the Belgian government to handle products from countries behind the Iron Curtain. The ideological divide of the Cold War and its endless number of rules and regulations severely restricted trade between East and West.

In all Eastern bloc countries, one had to deal with the State Trading Organizations. These vast monolithic organizations of state employees were responsible for buying all products needed by the particular industry they represented. They also sold all products produced by that industry. Laboratory and research chemicals in Prague were channeled through Chemapol, while Koospol dealt with the food industry.

Basic necessities usually sailed through customs. Czech hops, critical to the brewing of Belgian beer, or Belgian milk powder destined for Czechoslovak cooperatives, were considered as such. Similarly, the compounds sent by a laboratory of the Academy of Sciences in Prague did not raise any suspicions.

Holý selected three compounds representative of structural classes of his nucleoside analogues. A few months later an enthusiastic letter arrived in Prague. Unbelievable as it might have seemed, one of these synthetic compounds was antivirally active.

Seventeen more nucleoside analogues followed shortly after. By April 1977, Holý and De Clercq were ready to announce their first discoveries. Their article about DHPA, a broad-spectrum antiviral agent, in *Science* was delayed until May of the following year when the patent was finally filed.

The discovery of the compound changed Holý's life dramatically. Prolific as he might be, he was considered an introverted loner at the Institute in Prague. His colleagues did not seem to grasp his superb creativity. The cooperation with Erik De Clercq and the patent they had secured had suddenly given him a focus to channel his talents: developing antiviral drugs. The possibility that the DHPA compound could be developed into a marketable drug gave him instant authority. His image improved greatly and allowed him to speak directly with the managers of *Lachema*, the pharmaceutical plant that held the communist government's monopoly in Czechoslovakia.

Meanwhile, another compound arrived for screening at the Rega Institute. It was sent by Richard Walker's lab; the Birmingham compound was named BVDU and had been synthesized as an irradiation-sensitizing agent. Could it also have antiviral properties? At first, De Clercq did not think much of it. He tested it with a vaccinia virus assay and found nothing noteworthy. However, then his assistant tested it with a different virus and noticed the telltale activity with the cytopathogenicity method.[7] Human cells were grown in culture tubes. A dilution of the compound was added to one or more culture tubes, which were incubated overnight to allow time for the antiviral agent present to act on the cells. Then, all tubes were infected with the virus; if it was able to grow, it destroyed the cells. The areas of cell destruction indicated the extent of the cytopathic effect. If the antiviral in the dilution was working, the cells were protected so that no cythopathic effect was found.

"Compounds coming to life" meant that cells were protected! BVDU proved to be extremely active against herpes simplex.[8] Was the British BVDU compound similar to a discovery made at the Burroughs Wellcome plant in the United States?[9] Both compounds were tackling the herpes virus, a family of DNA viruses that have learned to play hide-and seek with immune cells. They cause latent infections that allow the viruses to remain in their hosts' bodies for life. Some herpes viruses are reasonably innocuous in healthy people and cause infections like cold sores, chickenpox and shingles. But in weaker and immune-compromised people, they can be deadly, while other members of the herpes family can cause cancer.[10] Richard Walker had been working along the same lines as Gertrude Elion at the Burroughs Wellcome facility but could not have possibly copied her. His compound had come alive in Erik's lab and received a new and thorough description of its mechanism in Leuven.

Had Richard Walker been inspired by a communist chemist from the GDR, an East German chemist, Peter Langen? Walker was also the editor of *Nucleic Acids Research* and Langen had written an obscure article in 1975 for his Journal. It was tucked away in the *supplement*, where it was not scrutinized by peer review. Langen had described the characteristics of the exact same compound but gave it the wrong chemical structure.[11] The description of the correct chemical structure earned Walker and De Clercq a joint patent for BVDU in 1978 and a moment in the limelight when Erik was invited to speak about this topic at a seminar in Prague.[12] A seminar, as fate would have it, organized by FEBS and chaired by the East German chemist, Peter Langen.[13]

Walker wanted to intensify his cooperation with the Rega Institute. He was rather annoyed that a great deal of Erik De Clercq's attention was still devoted to interferon because the induction of interferon was formally still his main mission at the Institute. Walker kept challenging him. Interferon had been studied for over twenty years and had not yielded any concrete benefit. "As a chemist, I do not even believe interferon exists as long as its chemical structure is not known. Prove it exists." That was a challenge Erik De Clercq could not leave unanswered. An invitation coming from the Ghent University would point the way.

Chapter VII
Breaking away from interferon

Fashions in science are as influential and nearly as mercurial as styles in dress.

— Arthur Kornberg

A molecule for all seasons

Could viruses cause cancer? The theory received a lot of credence in the early seventies and was a forceful driver for research in antiviral therapies and interferon in particular. The new focus on interferon as an antiviral, acting as an anti-cancer agent, was mainly the work of one woman. Mathilde Galland, a Swiss scientist, almost singlehandedly put interferon on the radar screens of both the NIH and the pharmaceutical industry.

Wherever she went, she was preceded by a legendary reputation. Her support for Jews in Palestine immediately after World War II led her as a young girl to ride her bicycle and collect guns in French villages to benefit the Irgun Underground. She converted to Judaism, married a Jewish scientist and moved to the newly created state of Israel in the 1950s. Their marriage, however, did not last. As a single mother and researcher at the Weizmann Institute in Rehovot, Mathilde caught the eye of an Institute's trustee, Arthur Krim, an American lawyer and president of United Artists, a leading film production studio. He was also involved in fundraising for the Democratic Party.

Her second marriage brought her to New York City in the whirlwind of Krim's world. It was populated with actors, movie stars, advertising icons as well as prominent politicians from the Democratic party. The famous birthday serenade from Marilyn Monroe for President John F. Kennedy in Madison Square Garden and the after-party in Krim's East Sixty-ninth Street townhouse placed her at the top of America's social circuit. Despite her new life as a leading society matron, Mathilde Krim remained faithful to her scientific interests and joined the research team of Sloan-Kettering Institute for Cancer Research.

She naturally gravitated into the orbit of another socialite and philanthropist, Mary Lasker. Her husband was the head of an advertising bureau and one of his ideas, the LSMFT *Lucky Strike Means Fine Tobacco* slogan had earned him millions of dollars. His vast wealth, acquired by promoting one of the greatest causes of cancer, now became instrumental in the fight against cancer. Mary Lasker and her husband created a philantropic foundation to support medical research. It became an essential base for the development of the National Cancer Institute within the NIH.

With her full-page advertisements in the New York Times, Lasker prodded President Nixon into action and was at the origin of his appeal for "a war on cancer." Lasker taught Krim how to successfully lobby the US Congress. Without hesitation, Krim branded the specter of viruses as a possible cause of cancer and greatly influenced the content of the National Cancer Act of 1971. Funding for cancer research was made a national priority. A closer investigation of interferon as a potential anti-tumor agent was part of the deal.

A three-day international workshop on Interferon in the Treatment of Cancer, organized by Krim at the Memorial Sloan Kettering Cancer Center in 1975, was memorable for many reasons. The interferon field until then was a rather fledgling community of scientists. She managed to bring together for the first time more than two hundred scientists, administrators from the NIH and representatives from the pharma industry. Together with her husband, Mathilde Krim invited the core of interferon researchers and her business connections into their historical home in Manhattan. Erik De Clercq was among the invitees, they were all overwhelmed by the lavish hospitality of the Krims.[1]

The participants were even more impressed when they witnessed how Mathilde Krim staged the scientific meeting as a media event. The conference was an important turning point whereby interferon was no longer considered solely for its antiviral properties but also for its tumor-fighting capacities. De Clercq lectured on the use of interferon in mice with cancer. He built on earlier studies done in France and contributed in his own way to Mathilde Krim's cancer awareness campaign.

The popularity of interferon received a temporary setback at another international conference at the Weizmann Institute in Rehovot in 1977.[2] The meeting with over two hundred participants was disrupted by some comments by one of the participants, Piet De Somer. He expressed his concern about side effects that had been observed following the injection of interferon in a Belgian patient; he did not mention his own arm, which had swollen up after he had injected himself with interferon. Some of the participants angrily responded that such negative news should not be made public just when interferon was finally gaining momentum. The incident was not widely publicized and in the end it was only a bump in the road for interferon's eventual ascendancy as a wonder drug.

By the time Mathilde Krim organized her second conference in 1979, the media hype was in full force. The written press, including *Time Magazine*, *Life*, *Reader's Digest*, *L'Express*, amplified by TV anchors like Walter Cronkite, raised completely unrealistic hopes in patients and their families. Krim marshaled funding both from the NIH and from institutions such as the American Cancer Society. Its massive purchases of interferon made in Finland underscored its new credibility. Tom Merigan, Erik De Clercq's former boss, was one of the recipients of this interferon purchase, which allowed him to perform the first lymphoma trials in Stanford. Pharmaceutical companies that until then had been mere passive spectators of the interferon scene became active patent chasers. Mathilde Krim remained influential all through the seventies and reinvented herself in the eighties by raising awareness for AIDS with AMFAR.

Interferon was no longer solely the domain of virologists and oncologists. It soon became a subject of study for immunologists once it became clear that, besides interferon, there were other proteins that were active in the body's immune responses. It was agreed in the mid-1970's that all proteins, including interferon, were called "cytokines" for their ability to spur the immune system in action.[3]

Cloning the interferon gene

The high profile of interferon was at the origin of the biotech craze. It propelled two start-ups onto Wall Street, Genentech and Biogen. Their share price soared on the stock exchange on the basis of promising research alone. Completely new dimensions in interferon research were riding on startling new developments in molecular biology, a technical revolution known as genetic engineering. It had been found that, based on the chemical composition of a particular messenger RNA, the corresponding DNA could be created. A strand of DNA could be cut and joined to a particle called a plasmid. This could then be inserted into bacterial cells where numerous copies could be made. The bacteria were acting like small factories for mass production!

Ten leading scientists created the company Biogen to commercially exploit the inventions and discoveries made in their respective laboratories. Their main goal was to clone human interferon.[4] Initially, all interferon studies were carried out in the Institute of Molecular Biology in Zürich, but later Walter Fiers, a professor at Ghent University and co-founder of Biogen, also became involved.

In order to isolate the genes, large quantities of human interferon were needed. No one in Fiers's laboratory nor in those of the other Biogen founders had any experience in running interferon assays. Some of the Biogen members, the Zürich group, worked with the Finnish institute of Kari Cantell and his leucocyte interferon created from white cell cultures. In March 1980, the Zurich Group became the first to clone human interferon from leucocyte cells, and renamed it alpha interferon.

Kari Cantell was making white cell interferon available to other researchers free of charge. However, Fiers needed interferon in much larger quantities for his experiments. He remembered the lecture Erik De Clercq had given at Ghent University about his work at the Rega Institute, inducing interferon in fibroblast cultures. He had induced human interferon synthetically according to Maurice Hilleman's method and obtained surprisingly large quantities of interferon. The Hilleman method was very practical to run assays but could not be used for therapies in human beings. This synthetic interferon was very toxic.[5]

With a few phone calls, Walter Fiers brought together actors from a range of institutions who in normal times would not even speak to one another. He linked them up in a unique Belgian cooperative chain that would become the first to clone the human interferon gene in fibroblast cultures. The starting point was the Rega Institute, where interferon was synthetically induced in fibroblast cells. The cells were then broken up and their nucleic acids extracted and taken to the Institut Pasteur in

Brussels. The Brussels team had the know-how to find the particular interferon messenger RNA that would be converted into DNA in Ghent University. The gene could only be identified indirectly through its ability to elicit interferon messenger RNA. To pick out the gene from within ten thousand bacterial clones was a formidable task. Once the gene was isolated and properly identified, interferon could be engineered. It was then sent to Erik De Clercq's lab to assay the results. The Belgian group confirmed their results in two articles in *Nature* in 1980, their interferon was renamed beta interferon. The race to clone interferon gamma was won by Genentech in October 1981.

Once the genes had been isolated, the chemical structure could be revealed and the way to the mass production of human interferon was wide open. The lack of "pure" interferon in sufficient quantities at a reasonable cost was no longer an obstacle to progress. In less than two years, more insight was gained than in the preceding twenty years since interferon's discovery. The pharma industry made plans to produce interferon to treat an assortment of malignancies such as hepatitis C, bladder cancer, multiple sclerosis, bird flu and SARS.

In a review entitled "Interferon: A Molecule for All Seasons," Erik De Clercq summed up the interferon story. What better way to leave the field than to give it an accolade? Now that he had met Walker's challenge to prove that interferon was not a dream, De Clercq could move on to explore new fields. He was now ready to concentrate on the world of nucleosides.

Chapter VIII
The first antiviral drugs

Most scientific discoveries belong to a continuous, collective process of exploration of nature rather than a series of individual explosions of imagination.

— Salvador Luria

NATO supports a nucleosides network

Erik De Clercq was introduced to the North Atlantic Treaty Organization and its *Advanced Study Institutes* in the most enjoyable way. He was selected to travel to the idyllic Greek island of Corfu together with about a hundred other investigators. The workshop tackled antiviral mechanisms and attracted a fine group of medical doctors and virologists.[1] De Clercq's old friend, David Shugar, was one of the stars of the meeting. Nobody paid any attention to the fact that his scientific homestead was based in Poland, not exactly a NATO country at the time. His Canadian passport was all he needed to gain a spot at the speaker's podium.

Another researcher from the National Institutes of Health, Robert Gallo, then barely forty years old, caught De Clercq's attention. He had plenty of nervous energy. Erik immediately felt he was a kindred spirit, one of the few people who shared his passion for retroviruses. A retrovirus with its unique enzyme was a hot topic for scientists in the early 1970s, but a few years later no longer seemed interesting. It was considered to be at the periphery of the grand questions of modern biology. Gallo, however, was determined to prove that retroviruses could disrupt not only animals but also humans. He was on a hunt to find at least one retrovirus that caused cancer in humans.[2]

Gallo was immediately interested in Erik's discovery of a substance that was active against the Moloney murine leukemia virus, an animal retrovirus. The substance could contain the replication of the retrovirus, but unfortunately had no effect on the cancer cells.[3] Its common name was suramin, a compound known since the 1920s and used in treatment of African sleeping sickness, a tropical disease caused by microscopic parasites.

Rather than dismissing De Clercq's findings, Gallo encouraged Erik to publish this story in his Journal *Cancer Letters*. Gallo's suggestion came as a total surprise. So far, no publisher had shown any interest in De Clercq's findings on polynucleotides and retroviruses. The reviewers claimed that its enzyme, reverse transcriptase, had no biological relevance. De Clercq's article appeared in one of the Journal's 1979 issues. Five years later, it would suddenly come to the fore when NIH scientists were desperately seeking a means to combat another retrovirus, the AIDS virus.

After their meeting in Göttingen, Richard Walker and Fritz Eckstein asked Erik De Clercq to join them in developing a nucleosides network. The three men felt it was important for scientists to step out of the laboratory, confront theories and exchange test results with people coming from different backgrounds. Rather than confining scientists to their field of expertise, their platform would bring together virologists, chemists, pharmacologists, clinicians, and representatives of the pharma industry.

Richard Walker brought his experience from the publishing world to the table. Fritz added his prestige and that of the Max Planck Institute, while Erik would take care of all administrative questions. With that, they started working on their common project. All they needed now was to find an attractive place in relaxed and pleasant surroundings for their meetings. And what could be better than the Italian countryside?

Searching for ways to fund these gatherings, De Clercq's experience in Corfu served as an inspiring model. NATO's scientific affairs division would be a perfect partner. It provided the funding and means to gather scientists from both sides of the Atlantic. For young scientists, it was truly a blessing to participate in such a forum. In the seventies and eighties, communications were still rudimentary. The personal computer and internet were not yet commonplace. The photocopier and fax machine were the only sophisticated devices available at that time.

The NATO administrators set strict conditions: The Advanced Study Institute (ASI) could only include scientists from NATO member countries; no country, not even the US, could be overrepresented with more than twenty participants. One truly remarkable rule was that the study courses had to take place over a period of at least ten days. It was believed that a minimum of ten days together was necessary to build the kind of lasting relationships that can serve as a cornerstone for productive science.

Walker, Eckstein and De Clercq worked diligently to select the 100-odd participants for their NATO-ASI. The first course took place in May 1979 in Sogesta, close to Urbino, in central Italy. They abided by the rules of inviting only scientists from NATO member countries, but decided to test the limits. Could they add a scientist from a communist country if they covered the costs? They did not explicitly ask and did not get a formal refusal either. So, they selected Peter Langen from East Germany, financed by FEBS, the Federation of European Biochemical Societies. The workshop carried a promising headline: *Nucleoside Analogues. Chemistry, Biology and Medical Applications.*[4] Richard Walker was going to be the master of ceremonies. He would introduce the speakers and weave a common narrative through their presentations.

Walker was driven by the deep-seated dissatisfaction with the way the work of chemists was treated. He felt that in the past few years many useful and potentially useful nucleoside analogues had been synthesized but little more had been done with them beyond a few perfunctory biological tests. He recognized there was a dearth of adequate knowledge about the available testing procedures.

In the proceedings of the first NATO ASI, Walker did not mince his words. He expressed his irritation over a lack of communication between the chemist, pharmacologist and the clinician. As a result, he believed that few compounds received the testing and evaluation they deserved. Another cause of Richard Walker's frustration was the fact that so much attention and research money was going to interferon and not enough to nucleosides research.[5] He was pleased to introduce a company that was an exception to the rule: Syntex, the first pharmaceutical company in Silicon Valley. It had made its fortune thanks to some blockbuster products like "the pill," the first effective oral contraceptive. Very early on its Institute of Molecular Biology explored the effects of nucleoside analogs on nucleic acids biosynthesis and cell growth.[6] Around 1970, the management of Syntex abolished its Institute of Molecular Biology. But John Moffatt, however, as the head of the new research department, kept the tradition and skills alive.[7] Two of his brightest co-workers, John Martin and William (Bill) Lee later used these skills to develop some of the best anti-HIV drugs that came on the market in the twenty-first century.

At Sogesta in May 1979, John Moffatt was one of the most captivating speakers. He was glowingly introduced by Richard Walker. He and Moffatt shared a common connection, a formidable mentor: the charismatic Indian-American chemist, Gobind Khorana who received the Nobel Prize in Medicine in 1968. John Moffatt was among the few graduate students ever trained by Khorana while Richard Walker was one of the many postdoctoral fellows coming from Khorana's orbit.[8] Their mentor's love of nucleoside and nucleotide analogs was infectious and permeated the atmosphere of the first NATO course. The sojourn into this medical frontier was an exciting experience for all participants and had them yearning for more.

Bringing antiviral therapy to the clinic

Now that Erik De Clercq had his boss's blessing to devote himself entirely to nucleosides research, he also had to suffer the consequences. Even though De Somer was now the rector of the university he still kept a close watch on his institute. Just about every day, Piet De Somer made the rounds in the Rega Institute, relentlessly asking the same question, "And? Did you find anything new, anything that could become a drug?".

A drug in De Somer's eyes meant something to try out on human beings, if needed, on himself. If the compound from Birmingham, UK was indeed active against one or two types of the herpes family, there was no better way to find out than to test it on his patients. Luck lurked just around the corner. An urgent phone call came from the university clinic begging for "something" to alleviate the pain of a nun. She was suffering with cancer and had also acquired herpes zoster, better known as "shingles." De Somer thought it was an ideal opportunity to try out the British-Belgian compound, BVDU. "If it works, it's another milestone in antiviral drug development, if it doesn't and worse, if the nun succumbs, you should not worry, she will go straight to heaven," he told De Clercq.

The nun survived and the shingles miraculously disappeared. A few weeks later, a prominent speechwriter of Piet De Somer, then rector of the University, also came down with shingles. Erik De Clercq was summoned again: "You better cure this man, I need him to write my speeches." Two more cases presented themselves shortly thereafter. All were cured in no time and Erik De Clercq celebrated his victory with an article in the British Medical Journal in 1980.

Looking to license the compound, he now had the necessary confidence to engage with Searle UK, as the company already had a working relationship with the scientists in Birmingham. Searle acquired the license. Everything appeared to be moving smoothly, BVDU was on its way to becoming a drug; the shock would come later.

De Clercq did not feel at ease doing experiments with patients, but he was emboldened by the articles written by an American researcher from New Orleans: Herbert Kaufman had changed the destiny of Bill Prusoff's compound, IDU.[9] He used it for the first time as an antiviral drug in the clinic. After administering IDU to treat herpes in the eyes of rabbits, he subsequently tried it on his patients. He was not hampered by any permits or restrictions imposed by the FDA which in the early sixties was not yet the strict watchdog it later became.

More than a decade later, the science community was awoken by Richard Whitley, a young professor-pediatrician, another pioneer in clinical trials of the first antiviral drugs. His trials at the University of Alabama were followed with bated breath. In contrast to Kaufman, he was very rigorous and kept the FDA informed about every step he undertook. In 1976, he reported on his trials with vidarabine which had been administered intravenously to patients suffering from herpes zoster (shingles). In 1979 and 1980 he received permission from the FDA to use acyclovir on babies with congenital herpes simplex infection. Whitley had singlehandedly given credence to acyclovir, the discovery that Gertrude "Trudy" Elion had described in December 1977. There was still a long and thorny road ahead for this compound. Once approved by the FDA, acyclovir would usher in a new era of antiviral drugs.

Many pharmaceutical companies were eager to poach Whitley as an advisor. Searle, then under the leadership of Donald Rumsfeld, was among them. The Californian company, Syntex, also was eager to get his advice. That is where Richard Whitley crossed the path of a young chemist, John Martin who was not yet thirty years old when he had synthesized gancyclovir, a compound of the same family as acyclovir. The two would develop a lifelong friendship.

Whitley's set of values and strong sense of social justice immediately appealed to Martin. He was inspired by Whitley's stories from Duke University about how he stood up against discrimination of African-Americans. Whitley became the compass for many of Martin's initiatives throughout his career; not only at Syntex but also when he moved to Bristol-Myers and later onto Gilead Sciences.

Chapter IX
AIDS emerges in the shadow of the Cold War

The importance of science as a tool of international diplomacy is not to be sneezed at. We scientists are extremely lucky to be able to slip into foreign cultures almost unnoticed, at every stage of our lives.

—Sir Tim Hunt

The East German connection

The Fifth International Congress of Virology was an imposing title for a meeting in a small room in the University of Strasbourg. There, Erik De Clercq became acquainted with Professor Hans Rosenthal, a virologist from East Berlin. The professor struck his walking cane against the table every time he disagreed with a speaker, which occurred rather often. An amused De Clercq made plans to visit Berlin shortly thereafter. The news that an East German state company had shown interest in Peter Langen's BVDU made it all the more urgent.

Driving his old Volvo, Erik De Clercq and his wife Lili prayed that no roadside assistance would be needed to travel across East German territory. One needed to have enough gasoline to make it all the way because stopping was not allowed. At Checkpoint Alpha in Helmstedt, as they left West Germany, the car was thoroughly checked. In search of political propaganda, soldiers rummaged through their luggage, overturned everything and looked underneath the car with a mirror device. They prodded a stick down the petrol tank and looked for the necessary equipment to change tires. The mileage and time of departure were relayed to checkpoint Bravo, the final destination. In order to reach Berlin, an island in the midst of the GDR, De Clercq and his wife had to drive along an eerie corridor, the Helmstedt-Berlin Autobahn, a highway that was protected by Volkspolizei with dogs to dissuade anybody from venturing outside the passageway. It had been twenty years since West Berlin was walled off and surrounded with minefields, barbed wires, tanks and armed border guards.

But as they approached the city, what a surprise! De Clercq and his wife, bracing, for drab and soulless quarters, found the Rosenthals, a warm and welcoming couple living in a stylish townhouse. They spent the weekend together exploring the "pearls" of the East. The next day, they drove to Dresden in the Rosenthal's East German car, a Trabant spouting exhaust fumes typical of the unrefined cheap petrol that was used in the East. Dresden was only partially rebuilt, its iconic cathedral still in ruins as if the Allied bombing raids of 1945 had just ended. Upon their return to East Berlin, it was mandatory to attend a concert of classical music in one of the beautifully decorated concert halls. The country's head of state, Erich Honecker, regularly attended televised events there as if to show the population how cultured their leaders were. On the third day of their stay, the police noticed that Erik and Lili had not formally registered themselves. This could have cost them some jail time, but Hans Rosenthal smoothed things out.

On another visit, Erik entered Berlin through its airport. Hans Rosenthal picked him up in his Trabant and drove him through West Berlin. With its flashing neon signs and bright lights, boulevards littered with cafés and restaurants, and music spilling out of theatres and cabarets, the city felt surreal. The odd East German car and its occupants received many scrutinizing looks from West Berliners as they drove by.

On their way to Checkpoint Charlie to crossover to the East, Erik began to feel somewhat uneasy but Rosenthal reassured him everything would be fine. The usually intimidating border guards seemed to have great respect and deference for this tall, imposing scientist. His Jewish background, and the fact that he had survived the Nazi persecutions doing forced labor during the war years, earned him special status. De Clercq's documents were stamped immediately, ahead of the long line of cars queuing to go through Checkpoint Charlie.

On yet another visit, Erik went to see Peter Langen in what was perhaps the largest science facility in East Germany, the Max-Delbrück-Centrum für Molekulare Medizin, in Berlin-Buch. Before meeting up with students for his lecture at ten in the morning, Langen cracked open up a bottle of brandy saying: "We need this here. How else can we survive?". Langen was shy and discreet and rather depressed by the communist regime. He nevertheless complied with all the rules. When De Clercq asked him to send some compounds for testing to Leuven, the answer was negative.

Langen confirmed that an East German state enterprise, Berlin Chemie, had plans to develop the East German version of BVDU. De Clercq reminded him that Birmingham UK and Leuven shared the patent as they were the first to detail the exact chemical structure of the compound. Langen steered him away from the topic with a gentle nudge: the German Democratic Republic did not recognize any patent rights from the West. De Clercq could not imagine then that eight years later the Wall would come down and that BVDU would become available to the whole of Germany, and that an Italian pharma company, Menarini, would buy out Berlin Chemie and market the drug under various brand names throughout the rest of Europe.

De Clercq's attention was focused on his next conference at the Robert Koch Institute of West Berlin. It occurred to him that the evening before, he had lectured at another Koch Institute, located within the "Charité," its walls still showing the bulletholes from the fight against Nazi troops during the war.[1] Were East and West vying for Robert Koch's legacy? He crossed the border from East to West-Berlin on the S-Bahn train in Friedrichstrasse taking him to the other side of the Wall. From the moment he arrived in the West, he was immediately engulfed by the atmosphere of freedom and exuberance.

Very few scientists in East Germany or the German Democratic Republic had the same leeway as their colleagues in other communist countries. Women, especially, seemed to be more suppressed than their Czech or Polish counterparts. In order to travel, even to another communist country, it was prudent to acquire the status of a married woman. This seems to be the path followed by a young physicist, Angela Kasner. She worked on quantum chemistry in the East German Academy of Sciences and married a physicist, Ulrich Merkel, in 1977. It was not a happy marriage but it enabled Angela Merkel, the future German Chancellor, to pursue her dreams. Frustrated by the lack of access to Western publications and scientists in East Berlin, she longed for Prague. In the Czech capital, at least according to rumors in the science community, access to anything that came from the West was much easier.[2]

Angela Merkel took up a series of postdoctoral projects in the Czech Academy of Sciences. Her professor of choice was Rudolf Zahradník, who had acquired some fame outside Czechoslovakia. Her new field, physical chemistry, happened to be the same as that of another of his followers, the East German researcher and chemist, Joachim Sauer. Under Zahradník's good offices, there was a happy confluence of Merkel and Sauer's intellectual interests. It also fit Angela Merkel's emotional state of mind. Her marriage was on the rocks and she found a new soul mate in Joachim Sauer. They later married in 1998. Prague with its many concert halls and opera houses was an ideal place for a romantic couple enamored with classical music.

Unfortunately, the Institute of physical chemistry, the Heyrovsky Institute, where Zahradník was working had remained bridled by the normalization atmosphere. He was often harassed and not allowed to supervise any postdocs, so he made sure his students and co-workers found refuge in other Institutes of the Academy.

His closest assistant, Zdenek Havlas, was parachuted into the IOCB, one of the institutes where life was more tolerable and where politics did not intervene in science. The fact that the IOCB had much better equipment than any of the other institutes added to its attractiveness. It also happened to be the institute where Tony Holý worked. Actually, Havlas had his office almost next-door to Holý's and it was only a matter of time before they became close friends. His stint as a postdoctoral fellow at the Cornell University with the renowned scientist, Roald Hoffmann, added to their mutual passion for chemistry.[3]

Angela Merkel needed access to computers for her work. The IOCB possessed highly-prized IBM computers, including one of the East-German copies that were off limits for most East Germans. The computers were monstrously big, occupying an entire room, and one needed to fabricate a punch card in order to make them work. Zdenek Havlas helped his fellow scientists to secure valuable computer time, if needed in Kladno, a thriving industrial town 30 minutes west of Prague, dubbed the "Czech Manchester" for its coalmines and related industries. The metal factories in Kladno allowed scientists to use their computers during the night.

All in all, Angela spent nine months in Prague over several years. Even though she did not speak Czech, she understood most of it thanks to her extensive knowledge of Russian which she had acquired as a teenager during her travels in the Soviet Union. She was quite fond of Havlas and his wife. One day she brought them a sewing machine from East Germany. Merkel was harshly interrogated on the Vindabona train she used to take from Berlin to Prague. Border guards harrangued her about illegally importing a sewing machine into Czechoslovakia. She finagled herself out of their questioning by insisting it was her own machine and that she needed it "to relax after her fatiguing research work."

On other occasions, she brought felt slippers from East Berlin for her friends in Prague. Czechs treasured felt slippers for walking around indoors, as it is an age-old Czech tradition to take ones shoes off when entering the home. The stock of felt in Czechslovakia was almost entirely siphoned off by the Soviets to manufacture cloth for the Red Army.[4]

When they were in private homes, away from indiscrete listeners in the lab, Sauer recounts that he, Merkel and Zahradník would try to lift each other's spirits. They speculated that this communist regime could not last forever.[5]

In Czechoslovakia, there were plenty of material things, but no food for the mind, no intellectual life except classical music. To see a play by Havel, forbidden in Czechoslovakia, you had to travel to Warsaw.

A joke in those days illustrated the situation. It is the story of an encounter between a Polish dog and a Czech dog at the border: The Polish dog asks the Czech dog why he wanted to come to Poland, "there is nothing to eat here." The Czech dog answered, "I know, but in Poland I can bark."

Angela Merkel's stay in Prague became well known thanks to an article that appeared in 1988, a year before the Berlin Wall came down. The article, authored by Merkel, Havlas and Zahradník became the most widely read article of the *Journal of the American Chemical Society* once Angela Merkel became the Chancellor of a reunited Germany.[6] Merkel's co-worker, Havlas, became Tony Holý's closest confidant. Once democracy returned to Czechoslovakia, he succeeded Holý as director of the IOCB. Rudolf Zahradník, became president of the Academy of Sciences and remained very close to Angela Merkel and Joachim Sauer. When Ms. Merkel and Mr. Sauer heard that the Czechs had no intentions of celebrating Zahradník on his 85th birthday, they honored him with a moving day-long event in the Academy of Sciences in Berlin. Chancellor Merkel was present for the duration of the event.[7]

The Cold War heats up

In 1983, little thought was given to reports about a strange illness that was spreading in the West. Behind the Iron Curtain, people felt far removed from this disease. Communist authorities portrayed it as another outgrowth of capitalism. The new illness seemed to attack mostly homosexual males. Its victims became susceptible to normally harmless common infections or afflictions, and suffered long and agonizing deaths.[8]

A rare skin cancer, Kaposi's sarcoma (KS), and a rare form of pneumonia, pneumocystis carinii pneumonia (PCP), generally seen only in severely immune-suppressed people were among some of those "opportunistic infections." Unusually large requests for pentamidine, a drug to treat pneumonia, alerted the Centers for Disease Control that something dangerous was going on. The disease specifically attacked T cells, the white blood cells that are crucial for the body's immune system. One could only speculate about the cause of this mysterious killer. Patients seemed to have high counts of alpha interferon which hinted at a viral infection. Physicians suspected that what made people sick was passing blood or other body fluids from one person to another.

At first, the Centers for Disease Control (CDC) called it "gay-related immune deficiency" or GRID. The gay revolution exploded during the eighties as young homosexuals sought to "liberate" themselves. They established their identity by engaging with multiple sex partners. Despite having plenty of evidence, the CDC was slow in admitting that heterosexual contact could transmit the illness as well. Instead, it issued a warning against the homosexual lifestyle.[9] It was only in August 1982 that GRID was aptly renamed "Acquired Immune Deficiency Syndrome" or AIDS.

By then, it was obvious that this disease did not only afflict gay men but also other people, such as intravenous drug users. Moreover, patients coming from Haiti, many of them infants and children, were showing indications that the disease could be passed from mothers to newborns. A small but growing percentage of infected victims were hemophiliacs, both men and women. As the media increased its coverage about transmission of AIDS through blood transfusions, the public wondered: Was the blood supply at risk?

Although the disease had been named and the syndrome identified, its cause was still not fully understood. Was AIDS a single syndrome with a common cause or was it a series of diseases? Was there a single infectious agent at work? Was it a virus and if so could it spread easily? Was the general population at risk?

The illness struck rich and poor alike. Age-old fears and superstitions, stemming from the history of plagues, reappeared. Some funeral parlors in the early days simply refused to deal with AIDS corpses. People with the new illness were stigmatized, in part due to homophobia and prejudice. Throughout part of the decade, U.S. Immigration Services did not allow AIDS patients into the country. Religious Christian groups brandished it "a punishment from God for promiscuous behavior."

When Richard Walker and Erik De Clercq organized their second NATO Study Institute in June 1983, AIDS had hardly received any attention in the nucleosides community. Les Arcs, a ski resort in the French Alps glittering in the spring, hosted some hundred-odd scientists, for the second ASI. Walker and De Clercq had convinced FEBS to extend sponsorships to scientists from communist countries. He was determined to get his Czech friend on board. Tony Holý would have a hard time explaining to Czech authorities why he was going to a NATO gathering. NATO was equated with the enemy. The East-West political climate was becoming tense as NATO prepared to deploy Cruise and Pershing-2 missiles to counter the threat posed by SS-20 missiles in the East. Espionage activities on both sides were intensifying. In Prague, anybody seeking to travel to the West would be viewed with suspicion by the communist rulers. Yet, an invitation from FEBS looked innocent enough, and Holý was given permission to attend the conference.

In the same manner, the Walker-De Clercq tandem helped sponsor their friends from East Berlin, Peter Langen and Hans Rosenthal. David Shugar was unable to join them because he had fallen into disgrace with the authorities in Poland. The first cracks in communism began to appear in the early eighties with the birth of the underground labor movement "Solidarity." All kinds of people were being questioned and harassed by the Polish authorities. Those with Jewish roots were seen as part of a "Zionist conspiracy." Shugar had to keep a low profile, as he knew his residency permit was at risk.

Thanks to the funding of some private companies, De Clercq and Walker were able to expand their reach. They contacted friends they had encountered the year before in Japan during a small, intimate workshop for a handful of nucleoside chemists in Kyoto. Their meetings were held in rooms with the typical tatami floor in the company of Walker's mentor, Gobind Khorana. He was admired for being the first to synthesize nucleic acids, for which he received the 1968 Nobel Prize in chemistry.

Another great addition at Les Arcs was one of De Clercq's co-workers, Shiro Shigeta, who had hosted him in Fukushima during a whirlwind tour of Japan. He was accompanied by a bright young student, Masanori Baba, who enjoyed the atmosphere so much that he asked to join De Clercq's laboratory. A few years later he would play a key role in his discoveries.

De Clercq had also selected a medical doctor, Richard Whitley, the pediatrician and professor from the University of Alabama, who was considered the smartest clinical researcher in antiviral therapies of his time. Whitley suggested they also invite John Martin, a chemist working at the Syntex company in Palo Alto. Martin had entered the galaxy of nucleoside chemists with his recent landmark achievement, gancyclovir, which would become a drug used to treat cytomegalovirus infections. But Whitley's suggestion was overlooked. Instead of John Martin, another Syntex colleague was chosen for the last slot reserved for American participants in the NATO Advanced Study Institute.[10]

It is not clear what participants found most memorable during those ten intense days of discussion in the French Alps. Spouses and children joined the scientists; it seemed to be a grand happy family. They clearly enjoyed the lighter moments, playing tennis and squash. The scientists were invigorated by the pure Alpine air and captivated by the new developments in their field.

Their German colleague, Harald zur Hausen, spoke about the discovery he had made identifying the causal link between certain types of the papilloma viruses and cervical cancer, which is the second most common tumor disease in women. Already in the seventies, zur Hausen had isolated these papilloma viruses, but now he had irrefutable proof of their connection with cancer. He had put all the elements in place to develop a vaccine.

There was much excitement over this news. All were convinced that one day zur Hausen would receive a Nobel Prize in Medicine. Twenty-five years later, in 2008, it finally happened. Zur Hausen shared the Nobel Prize with other scientists, Luc Montagnier and Françoise Barré-Sinoussi of the Pasteur Institute in France who isolated a virus found in the blood samples of an AIDS patient in 1983.

Revealing a retrovirus

It was Montagnier's assistant, Françoise Barré-Sinoussi, who had drawn his attention to the AIDS virus. It had all the telltale signs of a human retrovirus, a family of RNA viruses that had evolved with the unique capability of inflicting a lifelong infection of a cell while hiding from immune system attacks. Barré-Sinoussi had spent a few months in Robert Gallo's lab at the National Cancer Institute near Washington DC, where she became acquainted with the "human" retrovirus he had discovered. Montagnier, like Gallo, had been working on retroviruses since the early 1970's, but until then had only seen animal retroviruses. He was uncertain whether the virus he had isolated in a patient with AIDS was also the cause of the disease. He remained prudent and merely claimed that the virus seemed to be "associated" with the disease. He called it Lymphadenopathy-Associated virus or LAV. Could it be one of those "passenger viruses" that hitch a ride on a weakened body when the immune system cannot fight them off? This virus, although "associated" with the syndrome, could just as well be the "result" of the illness.

The speed and the way the disease was spreading had all the signs of an epidemic. Two cities in particular, New York and San Francisco, were under attack by the unknown killer. Calls to mobilize massive resources for research were left unanswered. Initially, the silence from the Reagan administration had been deafening. The press conference of Secretary of Health and Human Services Margret Heckler on April 23, 1984, came none too soon. The intense media coverage resembled the excitement of earlier times when important medical discoveries were made known to the public, like the polio vaccine or the cloning of an interferon gene.[11]

Unfortunately, Heckler's announcements regarding the cause of AIDS, the development of blood tests and the pursuit of a vaccine proved not very accurate. Her conference coincided with the publication of four papers in *Science* covering the results of Robert Gallo's laboratory at the National Cancer Institute; papers that explained in detail the behavior of the virus: how the retrovirus particles containing reverse transcriptase enzyme could convert their RNA into DNA once inside the cell.[12]

Gallo had been the first to identify retroviruses in humans, which are members of the HTLV or human T-cell leukemia viruses that trigger leukemia. It was an accomplishment he could genuinely be proud of. He went a step further than Montagnier and presented the virus as the probable cause of AIDS, but as a variant of the same HTLV viruses he had discovered in 1981 and 1982.[13] Gallo gave the name, HTLV-III, to the new virus he had isolated, but had used a cell culture that Montagnier had passed on to his lab during one of their many exchanges.

Gallo tried to clear up the confusion. The letter L in HTLV-III did not stand for leukemia but for lymphotropic. The accusation of fraudulently presenting Montagnier's virus samples as his own ignited a full blown transatlantic fight between the two scientists, who previously had been collaborating in an exemplary manner.

Traditionally, whoever made the discovery had the right to the name. But who should get the credit for the discovery of this virus? In order to resolve the dispute, an international naming committee was established under the chairmanship of the American scientist Harold Varmus, the head of the National Cancer Institute. By then it had become possible to study a virus through gene sequencing. It showed that LAV and HTLV-III were the exact same virus.

The committee rejected Gallo's claim that the virus was a variant of the HTLV-leukemia virus and proposed that the AIDS virus be called *Human Immunodeficiency Virus* or HIV instead. The committee also distinguished sub-categories of HIV. It recommended naming the more common type of HIV, the one in dispute, as HIV-1. The less common type seen in West Africa, a variety which Montagnier had discovered, could be designated as HIV-2.

The committee had done remarkable work, but was not able to settle the dispute. Gallo, unlike Montagnier, refused to go along with the May

1986 letter the committee sent to *Nature* which introduced the new name.

The media and most researchers followed the World Health Organization's suggestion to use both names until the imbroglio was solved. It took an agreement signed by President Ronald Reagan and French Prime Minister Jacques Chirac in 1987 to formalize the naming of HIV and to share patent rights to the technology for detecting infection of the virus.[14]

Meanwhile, at the NIH, the pressure mounted to find "something to fight AIDS." Sam Broder of the NCI had received permission to set up an AIDS drug initiative but did not get a significant budget nor any additional staff. His small team was quickly overwhelmed. Who could help fight this retrovirus? Gallo remembered Erik De Clercq's article that he had published in one of his *Cancer letters* of 1979, and recommended testing the effects of suramin on this virus. It produced a remarkable result. Suramin did not kill infected cells but blocked viral replication.

Broder called Erik De Clercq in October 1984 to congratulate him. In its latest issue, *Science* had accepted an article about suramin as an option to conduct human trials. Broder's Japanese post-doctoral fellow, Hiroaki Mitsuya, co-authored the article with Gallo. They extensively referenced De Clercq's findings. It pleased De Clercq tremendously, but notwithstanding the prestigious article, AIDS remained far from his mind in 1984.

He had just returned from the British branch of Searle in High Wycombe, where he had to cope with the news that the company no longer was interested in BVDU. It was not clear what prompted Searle's decision to cut its ties with both Walker and De Clercq. Was this sudden disaffection due to the new CEO, Donald Rumsfeld, former secretary of defense under the Ford administration? Rumsfeld wanted to streamline the company by shedding a number of global subsidiaries in order to focus on its core business.

The disastrous reports about a Japanese trial with BVaraU, a drug with close resemblance to BVDU, would later compromise the British-Belgian compound even further. The Japanese did not realize that their drug was incompatible with an anti-aging supplement, fluoro-uracil, that is very popular among adults there. Administering their drug to people with herpes had caused more than eleven deaths. Because of the close similarity in name, the two drugs, BVDU and BVaraU, created confusion; they were forever conflated with one another in the Anglo-Saxon world.

In October 1984, the media were more interested in Belgian scientists from the Institute for Tropical Medicine in Antwerp. Many of the patients with AIDS in French and Belgian hospitals came from central Africa. Once the hypothesis of an Africa link was triggered, Peter Piot's expertise was in demand because of his previous experience with viruses originating in Africa. He had been a member of an investigative team in the 1970s that found the reservoir of the Ebola virus. It allowed the Institute in Antwerp to isolate the virus from the infected serum of a missionary nun. The virus was then more thoroughly analyzed at the Centers of Disease Control and Prevention, which named the virus after the Ebola river.[15]

A CDC mission that went to Zaïre along with Peter Piot's group came back with devastating stories.[16] AIDS in Africa was not a so-called gay disease because the virus equally infected men and women. It was most prevalent, not among the poor, but among the educated elite who often had more sex partners than the average person.[17]

These reports were instrumental to universalize the disease in the western world and lay the groundwork for an international AIDS conference in Brussels in 1985. A few months after the Institute in Antwerp had introduced the famous ELISA tests in Belgium, the conference encouraged several African governments to organize blood screenings and create national committees that would consolidate information about the epidemic.[18]

Many Africans, however, felt they were being blamed for the epidemic. The suggestion that the AIDS virus originated from African monkeys was a particularly insulting form of racism. Many claimed that the virus had evolved in the West and was introduced in Africa by visitors, United Nations soldiers or foreign businessmen. Others asserted that the virus had been released into the native population through the distribution of tainted polio vaccines. A few leaders suspected a form of biological warfare, spread by European governments in order to cripple the independence of their former colonies. Public denial became a formidable obstacle in dealing with the epidemic in Africa.[19]

The representatives from 50 African nations, present at the Brussels conference, issued a statement saying there was no conclusive evidence that AIDS originated in Africa.

Chapter X
From passivity to action

The world of science may be the only existing participatory democracy. Science is an immensely supportive activity. Its support is both intellectual—the sharing of knowledge—and emotional—the sharing of purpose.

— Salvador Luria

A pivotal year

On a breezy day in March 1985, while making his rounds of the academic research institutions, Julius Vida, a licensing director from Bristol-Myers, appeared in De Clercq's office and asked if there is any product he would like to develop together with the American drug maker. He was a most agreeable man who knew how to impress De Clercq with his soft spot for chemistry. Vida had studied with the renowned scientist, Robert Woodward, at Harvard who was considered the most artful of master chemists in his era. Long before he received the Nobel Prize in Chemistry he had become a cult figure among scientists, even his idiosyncrasies, like his fixation with the color blue, were legendary.

With his old-world charm, Julius Vida belonged to the more sophisticated kind of Americans. He introduced De Clercq to top class restaurants in Brussels where cuisine and fine wines fused together like sublime chemistry, places where a young university professor would not often set foot. But on one of their outings, De Clercq accidentally said too much. He spoke of a new class of antiviral compounds and had to hold his tongue so as not to give away any more details until the compounds were properly patented and enshrined in a publication. "Come back next year," he told Vida, who immediately responded with an invitation to visit Bristol-Myers in the United States.

The discovery of a new class of antiviral compounds, the acyclic nucleoside phosphonates, was one of De Clercq's most thrilling experiences. Holý's compounds, HPMPA and PMEA, had come to life in the assay systems that his first Japanese postdoctoral fellow, Takashi Sakuma, had introduced in Leuven.[1]

De Clercq had been working with Holý unabatedly for almost a decade. They seldom saw each other, except for a few international conferences like the FEBS meeting in 1978 and the intimate workshop in Kyoto, Japan with seven other chemists and the Nobel Prize winner Khorana, in 1982. Most of their collaborative work was done over the phone or by correspondence. This time, however, he had to share his joy with Tony Holý in person. He accepted to act as chairman of a symposium on virology in Bechyne Castle not far from Prague. It was a good excuse to visit his friend at the IOCB and celebrate their invention in one of Holý's favorite restaurants: the century-old *Red Wheel*, near the convent of St. Agnes, the patron saint of Bohemia. It was also another opportunity to bring compounds to Leuven; his coat stuffed with plenty of new vials.

It was precisely in these happy times that De Clercq was struck the most dreadful blow he had ever experienced, with the sudden and unexpected news that his boss, Piet De Somer, had died. The abrupt loss of a legend left him and everybody else in Leuven in a profound state of shock. De Somer looked so vigorous not long before when he was hosting the Polish pope, John-Paul II, in Leuven. Never before had a pope visited this University, the oldest of all Catholic universities in the world.[2]

A few weeks before De Somer died, the United States beckoned De Clercq with travels to Bristol-Myers, a lecture at the US Army Medical research facility in Fort Detrick, Maryland and an urgent invitation from Sam Broder at the National Cancer Institute. At the NCI, a group of about fifty scientists with only a few Europeans, discussed whether there was an agent which could be effective against replication of the AIDS virus.

Sam Broder did not say much about the human trials the NCI had been conducting with suramin. The very first drug to act against a retrovirus showed promise in the lab. When tested on patients, the decrease in viral load was impressive indeed but the side effects from the weekly injections were just too toxic. Broder shared his consternation over the pharmaceutical companies in the US. Not a single one was interested in looking for a drug against AIDS! He had called all of them, from the largest to the smallest start-up, but all claimed there was no market.

The only firm with a different attitude was the North Carolina-based company, Burroughs Wellcome. Even though Wellcome did not want to work with retrovirus samples, the company had given the NCI some promising compounds to analyse. One of them, AZT, could become a drug against HIV. The representative from Wellcome, present in the room did not want to divulge too much information: "I can only say it is a nucleoside analog, but it is not acyclovir." This came as a shock to De Clercq. He had never imagined a nucleoside analog could be active against a retrovirus. So far, he had only used nucleosides with DNA viruses of the herpes family.

The Burroughs Wellcome compound, AZT, was revealed in greater detail a few months later, in the October edition of the Proceedings of the National Academy of Sciences (PNAS). De Clercq raced excitedly over to one of his younger colleagues in the department of medicinal chemistry, Piet Herdewijn, waving a copy of the article. He kept trying to convince the younger chemists in the Rega Institute to work on nucleosides. AZT had been synthesized in 1964 as an anti-cancer agent by Jerome Horwitz of the Detroit Cancer Center, but was not potent enough to become a cancer drug. It was nevertheless acquired by Burroughs Wellcome and kept on the shelf. Thanks to his previous association with the company, Sam Broder remembered the compound when he was at the NCI.[3] Horwitz had synthesized other dideonucleosides: d4T (stavudine), ddI and ddC. Could they also become drugs against HIV?

When De Clercq asked Herdewijn whether he could produce a similar compound as AZT, he received an answer within a month. By November 1985, Herdewijn had synthesized d4T, but did not know yet whether it was active against HIV. To assay this compound there was only one option: sending it to the National Cancer Institute lab in Washington.

A fortuitous coincidence: De Clercq's assistant, Jan Balzarini, was about to take up residency for his year-long sabbatical at the NCI to study the workings of HIV assays. He could test Herdewijn's d4T, it complemented perfectly Mitsuya and Broder's program as they had obtained the other dideonucleosides ddI and ddC, synthesized by Horwitz in the sixties.

The Christmas season was just around the corner. An eventful year filled with both sorrow and promise was coming to an end. After De Somer's death in June, Erik De Clercq was named the Head of the Rega Foundation, a legal entity that managed the funds following the break-up of Rega and RIT. At the Rega Institute itself, the succession as the head of the institute was proving an impossible task. All five of De Somer's assistants, except one, felt a calling to become his successor. As there was no consensus, De Clercq, who had absolutely no interest, was designated not only as the Head of the Institute but also as Director of the Microbiology Department in the medical faculty. The youngest member of De Somer's inner circle who had never been in charge of a team larger than five people was suddenly entrusted with the unique legacy of his boss.

De Clercq was still mourning the passing of Piet De Somer, and tried to overcome his grief by writing an article for *Nature* about the phosphonates that Holý had prepared and the antiviral activity that had been discovered in Leuven. If everything worked well, it could be published before the end of the following year. At the end of his article, he added a suggestion: "One day these compounds should be investigated for their capacity to fight AIDS!".[4]

By the end of 1985, after four years of indifference, the general public in the US and the rest of the world was slowly becoming aware of AIDS. The disease was given a face after Hollywood heartthrob, Rock Hudson, sought treatment at the American hospital in Paris. He had joined hundreds of other Americans who had flocked to the hospital hoping to receive Montagnier's experimental drug HPA-23.[5] The announcement he had AIDS was a bombshell.

Weeks before Hudson died, President Reagan was asked during a press conference whether he would support a massive government research program against AIDS like the one that Nixon launched against cancer? He surprised everybody when he pronounced the word "AIDS" for the first time and assured it would be a top priority for the US government: the 1986 budget would earmark half a billion dollars for research on AIDS! However, once media attention receded, his proposal was reduced by twenty-two percent in Congress and Senator Jesse Helms started adding his notorious amendments to every appropriations bill, limiting research and prevention of AIDS in the US.[6]

A triangular collaboration is set in motion

The first cold weather of 1986 brought Julius Vida back to the Rega Institute. He was eager to learn more about the new class of antiviral compounds. However, De Clercq kept tight-lipped as his article in *Nature* had not yet been published. He did not feel like sharing any information and only mentioned that he had been working with a chemist in Prague. He was certain that Vida, however, would be too discouraged to go anywhere behind the Iron curtain, as was the case with most of his interlocutors. Julius Vida, however, was of Hungarian origin, and before emigrating to the US, had studied in Budapest and often traveled to Prague. Instead he bubbled over with excitement, and immediately made travel plans to visit Prague.

Vida loved the contrarian nature of Czechs. Even Prague's river, the Moldau, expressed contrarianism. The river cleaves the city in two, not unlike the Danube between Buda and Pest. All rivers in Central Europe flow to the Black Sea, whilst the Moldau heads in the opposite direction and joins the North Sea. Vida was saddened to see how Communism and Soviet rapacity had brought this country to its knees. He was not deterred by the air of neglect that lingered over the city. The baroque churches and Italianate palaces with marble staircases, the ornamental ceilings peeling with broken plaster were just waiting to spring back to life.

The meeting with Tony Holý was filled with expectations. The suave pharma representative, polished from Harvard, and passionate about nucleosides and nucleotides, found a laboratory thriving under adversity, a hidden treasure grove. The news that a representative of an American company came to see Holý at the IOCB generated a lot of nervous energy around him. *Inventia*, the office for defending intellectual property rights, was immediately put on high alert. In the meeting, Julius Vida got straight to the point and asked if Bristol-Myers could examine Holý's compounds.

Vida realized early on that Holý and De Clercq were like two sides of the same coin. He sensed that De Clercq treasured his first big discovery, BVDU. In order to coax his willingness to cooperate, Vida arranged for a meeting with one of the big bosses in New York. The Vice President of research, Giulio Vita, reigned supreme over the posh Manhattan headquarters at 345, Park Avenue. The elegant building and all the hushed formalities surrounding the VP duly impressed the Flemish scientist.

De Clercq prepared a presentation on BVDU, a compound similar to what Gertrude Elion had discovered, but promised more activity in a wider range of herpes viruses. Giulio Vita had a short attention span, however. The instant he learned that Searle had already produced the drug and returned the license, the meeting was over. "I am not interested in violated virgins," he grumbled. The talk with Giulio Vita would have been the definite nail in the coffin of BVDU if scientists in East Berlin had not salvaged the drug. After the Berlin Wall came down, it became a popular drug all over Europe.[7]

In 1986, however, it seemed as if De Clercq went from one failure to the next. In Washington, his assistant, Jan Balzarini, had tested d4T against the AIDS virus in Sam Broder's lab. He had not found any noteworthy activity however. The ATH 8 cells used in Sam Broder's lab were possibly not reactive enough. Yet, it was in the same cell line that activity of nucleosides of the same family like ddI and ddC was detected. The mystery was never solved.

The only silver lining in those days was the visit in Leuven of a junior Bristol-Myers executive: The Associate Director of Anti-Infective Chemistry, John Martin. Though unassuming, the thirty-five year old scientist had already acquired quite some feathers in his cap. De Clercq's lecture at Bristol-Myers in Syracuse had piqued his curiosity. His colleague, Julius Vida, only added grist to the mill with his gushing comments about Leuven and Prague. Martin could not contain his excitement knowing that De Clercq and Holý were working on phosphonates. He wanted to know everything about it.

Before moving to Bristol-Myers, John Martin had synthesized the new antiviral, gancyclovir. While he was at Syntex, he had also synthesized phosphonates when nobody was interested in this field. The boss of research, John Moffatt, who had explored phosphonates in the late sixties was pleased, but the company never wanted to investigate these further. When he left the Californian company in 1984, Martin was not able to take the results of his experiments with him since the intellectual property belonged to Syntex. It had been a gnawing frustration.

John Martin's arrival in Belgium, just at the start of a sacrosanct holiday, delayed his encounter with De Clercq. He used the May 1 weekend to travel to Bruges and soak up the age old traditions permeating that part of the world. He had read many articles written by the newly minted Director of the Rega Institute and knew him from the times he had visited the West Coast. Their first encounter took place at the Syntex conference hosted by Thomas Merigan in 1981, where De Clercq gave a lecture on "Selective Antiviral Drugs." John Martin vividly remembers their second meeting at a Symposium on Antiviral Agents organized in Seattle by the American Chemical Society in March 1983. It was a great honor for both Martin and De Clercq to be chosen as two of the main speakers for the prestigious event.

When Martin finally sat down with De Clercq, he was pleasantly reminded of how well the virologist could communicate in the universal language of chemistry and introduce him to Belgian beers at the same time. It reassured John Martin that he was on the right track in his renewed pursuit of the phosphonates.

Martin subsequently flew to Prague to visit Tony Holý at the IOCB. They had met each other before at several international conferences but never had the chance to deepen their acquaintance. He found Holý sitting at a small desk with a pile of photocopies and reprints 1.5 meters high, which if it were to fall over would surely have knocked him out and ruined the experiment as well. Most of the space in his laboratory was taken up by his technician, a tall imposing woman several years his senior. He was astonished to see how Holý and his assistant were working with homemade reagents without any protection, no laminar flow hoods, nothing. His lab was as plain as a kitchen.

Holý overcame his innate suspicion and distrust and showed his American visitor the center of Prague. They walked around the old town with its narrow streets and cobblestone alleys. It was a great feeling to cross the fourteenth century Charles Bridge and walk up the hill to the castle. Martin sensed the city's hidden beauty behind the walls polluted with the soot of brown coal. John Martin's acquaintance with Prague and his immediate trust in Tony Holý, cemented the *Cold War triangle,* the triangular collaboration set in motion by Julius Vida.

Upon his return in Connecticut, John Martin was approached by a very charming female CIA officer. "Could he debrief her about his contacts with Holý?" she asked, fluttering her eyelashes. He had to disappoint her. He declined and strongly advised her not to enlist Holý as an asset in Czechoslovakia. Politics and science don't mix!

Launching an AIDS laboratory in Leuven: the story of d4T

De Clercq was wondering what could have gone wrong. He was tinkering with all the steps necessary to assay Piet Herdewijn's d4T and kicked himself for not being able to carry out the test himself. Just then, one of the visiting fellows, an assistant professor in ophtalmology from the University of Pennsylvania named Herbert Blough, ambled into his office. He was on a two-month mission in Europe, combining work at the Rega Institute with research at the Pasteur Institute in Paris and wanted to test some compounds to see whether they could be active against AIDS. Almost as an afterthought, he mentioned the vials with HIV viruses he carried, courtesy of Luc Montagnier.

De Clercq felt quite helpless since his only assistant was in Washington. There was just a postdoctoral fellow from Japan working with him who was busy testing Holý's compounds and a student in pharmacology, Rudi Pauwels. He was about to leave the Rega Institute as his two-year contract was coming to an end; an internship in a nearby pharma company took up all of his time. De Clercq nevertheless called on Rudi, kindling his curiosity:

> Could he interrupt his work to help a professor from Philadelphia? He has found some compounds that could work against AIDS and wants to test them against the virus samples he brought with him!

Rudi jumped on the occasion, eager to help. He came back to the Institute immediately. He showed Blough where to change into a protective suit to enter a lab that, at best, had minimal safety levels. All technicians fled when they heard the professor possessed vials with the AIDS-virus. One technician, named the "mother of the laboratory," remained in the area and told Rudi Pauwels what to do in order to prepare the cells. Just as Blough was going to open the vial, he suddenly remembered he had to make an urgent phone call and began acting very nervous. The fear of the virus was too strong. He pretended he had to urgently leave for Paris.

"Could Rudi Pauwels continue the testing?" he meekly asked. "Step in a protective suit, open the vial, and do the test." Rudi did just that and, a few days later, found a very mild action against HIV. An article was prepared for publication, Herbert Blough was never seen again, but AIDS had made its entry as the new research topic for the Rega Institute![8]

Rudi had tasted the thrilling experience of working with the AIDS virus. His hunter's instincts got the upper hand. It felt like playing Russian roulette. He was barely twenty-five years old, but he considered the fight against HIV a military mission. He was now sure he wanted to work on one thing: installing an AIDS lab at the Rega Institute.

As the AIDS laboratory was being set up, everybody objected to bringing the virus into his or her side of the building. Negotiating with his colleagues for more space was not De Clercq's forte. Rudi Pauwels had noticed that the university hospital was about to move to another location. The emergency department had already liberated some space and the basement of the clinic linked to the Rega institute by an underground corridor. Even though there were no windows, it seemed like a perfect place to start working with the virus.

Rudi cajoled Jan Desmyter, head of clinical virology lab at the university clinic to step in. It was a happy confluence of minds. Desmyter had been thinking of bringing the virus to Leuven for research ever since he met Luc Montagnier. Thanks to his Parisian contacts, the institute had ready access to virus samples, but the tricky part was to multiply them.

Rudi Pauwels worked day and night to set up an AIDS assay system. He collected all kinds of cell-lines to see where HIV multiplied best. For HIV to infect cells, it must dock with a receptor that sits on the surface of those cells. Pauwels managed to acquire Japanese leucocyte cells that had been weakened through pre-infection with the leukemia virus and had plenty of CD4 cells. The MT2 and MT4 cells from Naoki Yamamoto's lab in Tokyo were a delight. They were unusually sensitive to HIV. Ironically, Yamamoto's lab was at that time also developing the d4T compound.

Rudi did not only have a nose for finding the right cells but he also invented a completely new system to assay HIV. It was much simpler to use than the traditional assays that involved measuring the cythopathogenicity and required a double check under a special microscope, a cumbersome device. With Rudi's system one could detect HIV activity with the naked eye, on the basis of colors: brown for infection, light yellow for protection. He had managed to automate the assay by introducing robots which he had fabricated in his garage at home. His assay system became the talk of the science community. The article that was published later in the *Journal of Virological Methods* received a gold star in the citation index. For over twenty years, it was considered the best system around and used in just about every lab in the world!

At first, no technician wanted to work with Rudi Pauwels for fear of being infected by the virus, but help was on the way. A new Japanese post-doctoral fellow, Masanori Baba, arrived in Leuven. He was the student of one of De Clercq's best Japanese friends who came to his NATO conference at Les Arcs. There and then he asked to join De Clercq's lab. Applying for financial aid took a while, but he obtained the very prestigious *Fellowship from the Japan Society for the Promotion of Science* (JSPS). Baba had remarkable skills and brought with him two more assay systems to the Rega Institute: a test for adenoviruses and a new test for varicella-zoster viruses.

As one of his first assignments, in August 1986, he assayed Piet Herdewijn's d4T compound. Baba found immediately that the compound was very active against HIV. It was a complete turnaround from the testing that was done eight months earlier at the NCI lab! This time, the patent lawyer in the Netherlands was contacted immediately. As to a publication, De Clercq looked for a journal that would publish swiftly and not wait an entire year for all peer reviews to arrive. He found a journal in Madrid that promised to publish within three months.[9]

De Clercq asked Bristol-Myers whether they would be interested in the drug. Julius Vida confirmed that they were. Actually, they had already acquired another compound of that same family, ddI and planned to offer this drug for "compassionate use" to the gay community. They also knew there was something in the works in Bill Prusoff's lab in Yale. Bristol-Myers would go with whoever would get the patent first. The race between the two universities was on.

The patent lawyer in the Netherlands took an unusually long time. There was no reaction, even though the article had been sent to the publisher. A few months later, in late 1986, De Clercq's friend at Yale, Bill Prusoff, filed his patent. Filing meant he was first in line to be awarded intellectual property ownership. It sent shockwaves throughout the Rega Institute in Leuven. Did De Clercq influence Prusoff when he gave a lecture at Yale in May?

As was the case with AZT and the other dideoxynucleosides synthesized by Jerome Horwitz, the chemical substance of d4T was already described in literature, therefore the patent application could not be for its chemical composition but rather for a method of use. Prusoff had filed a patent after discovering that d4T was not toxic to human cells in cell cultures, but he did not possess any viruses in his lab, let alone HIV. How could he prove the antiviral activity of the compound? One of Prusoff's postdoctoral fellows, Raymond Schinazi, who had just moved to Emory University School of Medicine to study viruses and immunology with his uncle, a well-known herpes researcher, held the key. He proved in 1987 that d4T was indeed active against HIV.[10]

Yale had filed the patent first, but Leuven felt they were the first to invent. Could the Rega Institute prove this in court? Would De Clercq start a battle with Yale, and what is more, against one of his best friends, Bill Prusoff? Yet, the goal posts were moving. Demonstration of antiviral activity in cell culture was not enough to support a patent. At that time, the patent office required proof of antiviral activity in human clinical studies. That ruling was not taken into consideration in Leuven. All hopes were still vested in the Dutch patent lawyer while Bristol-Myers was already introducing d4T into human clinical trials.

Although there were never any ill feelings between Prusoff and De Clercq, John Martin put an end to the bickering by inviting them both as co-authors in his article that appeared in "Antiviral Research" in 2010.[11]

Chapter XI
First attempts to halt the epidemic

One person can make a difference and everyone should try.
— John F. Kennedy

Two irons in the fire: Bristol-Myers and Janssen

De Clercq and Holý's article about the acyclic nucleoside phosphonates appeared in *Nature* on October 2, 1986. It was a follow-up to one of their first articles. This time it was even more authoritative: a new class of antiviral compounds was born. Their antiviral activity was captured in Sakuma's assays and a friendly opthalmologue in Leuven had tested them on eye infections in rabbits.

The authors received praise from all sides and were courted even more intensely by Bristol-Myers. Talks to acquire a license became more pressing. In order to study the compounds, Bristol-Myers had to copy a few samples. Holý provided them with guidance while De Clercq tested and analyzed the copies. They enjoyed all the niceties that were thoughtfully arranged for them on their visits to the new research facilities in Wallingford, Connecticut: limos waiting at the airport and relaxation time to practice his favorite sport, squash. De Clercq nevertheless remained wary after the d4T episode. But John Martin was so enthusiastic and prophesized: "You will develop something much stronger than d4T. You have phosphonates!".

As for Tony Holý, every time he visited Wallingford, he went on a shopping spree at the hardware stores. He was always on the lookout for the latest gadgets and tools. Bristol-Myers executives knew Holý was fond of playing pool, so a billiard table was reserved for him and his assistant. Law enforcement officials kept circling around John Martin. They wanted to know what Holý was up to in the United States. They were particularly worried about him taking pictures everywhere with his little camera.

Martin's mundane responses to their questions, "This time he bought a fancy screwdriver," exasperated them.

Meanwhile the Rega Institute was expanding Rudi's AIDS lab with the tremendous support of Jan Desmyter who acted as an all-in-one spokesman, broker and promoter. The activity at the Rega Institute sparked the interest of the legendary Belgian drug maker, Paul Janssen. He had been a friend of Piet De Somer. Both had inspired each other with their ever expanding curiosity and an unquenchable thirst for novelties. Both relentlessly pursued their staff with the same question "Is there anything new to report?".[1] They had an unwritten gentleman's agreement not to approach each other's collaborators. Now that De Somer had passed away, "Dr. Paul," as he was affectionately called by his assistants, invited Erik De Clercq to his stronghold not far from Antwerp.

Beerse was the small town that hosted the pharma installations and several of the office buildings where Paul Janssen had started his company in 1956. He merged it five years later with Johnson & Johnson to allow his laboratory potential to grow.[2] In less than twenty years time he had motivated some 1,300 young people to join his company. He hired them not on the basis of their school or academic degrees but on the basis of their ability to memorize. Next he helped his newly hired men and women expand their knowledge in one or another field, sending them to academic courses and asking them to focus on a specific subject until they became expert medicinal researchers. Even though Janssen pharmaceutical belonged to an American group, it was and still is the pride of Belgium.

By the time De Clercq went to visit Paul Janssen on a grey November day in 1986, his company had already invented more than fifty drugs, five of which had been posted on the WHO's list of essential medicines. Janssen had also acquired world fame with the opening of his plant in China, being the first western pharmaceutical company to set up a factory in the People's Republic of China.[3]

Janssen was unable to bring HIV inside his facilities, at that time the virus was still considered too dangerous and AIDS was surrounded by all kinds of taboos. The general public placed the blame for the spread of AIDS squarely on the gay community and anger mounted as more people died. However, Janssen, who had travelled in Africa knew better. He was shocked and obsessed by what he had seen there.

Erik De Clercq knew Janssen only superficially at that time. He had met him socially at functions organized by De Somer, where Janssen's extrovert wife was always the life of the party. People loved to be in Janssen's presence, his mastery of any subject from history to architecture to science to linguistics was a pleasure for the mind. It was a great experience for De Clercq to sit down with him for lunch at the local restaurant. They must have talked about everything under the sun because eight hours later they were still sitting in the restaurant. By dinner time they had made up their minds: they were going to work together!

Once Janssen and De Clercq decided to join forces, it took the university establishment another six months to formalize their agreement. Janssen would finance fellows and the Rega institute would examine a library of 600 Janssen compounds as a starting point. It was a splendid boost for the brand new laboratory that had been set up by a student, Rudi Pauwels, who had not even started work on a Ph.D. Yet, AIDS research was now firmly established in the Rega Institute.

When Jan Balzarini returned to Leuven, after his sabbatical year at the National Cancer Institute (NCI) in Washington, his mission was not entirely overtaken. He introduced the system he had gained experience with in Washington on a different floor than the basement laboratory Rudi Pauwels had assembled. They worked in totally separate ways, technicians who learned to work with one system did not want to adopt another way of following the protocols. The Rega Institute now effectively possessed two AIDS laboratories: one based on Rudi Pauwels's system that soon would acquire world fame and another based on the Broder and Mitsuya method of the NCI.

The NCI had grown in stature not in the least because of Sam Broder's accomplishment. He had successfully steered AZT (azidothymidine) through regulatory procedures from test-tube to patients in a record-breaking 19 months. The drug was given to patients for the first time in July 1985, the phase two clinical trials starting in 1986 as a double blind study had to be aborted. It was almost immediately obvious that many more people died on the placebo than on the drug. The effects were so immediate and protests in the gay community were so violent that trials had to move straight to the final phase.

Taking stock after AZT

The elation following the FDA's approval of AZT did not last long. It soon became obvious that the drug had plenty of side effects and only allowed life to be prolonged by a year.

It was nevertheless an occasion for President Reagan to give his first (and only) public speech on the subject of AIDS. The day after he and French Prime Minister Chirac announced that French and the American scientists would share credit for the discovery of the virus, he addressed a conference of medical doctors in Philadelphia in April 1987. Reagan remarked on how American scientists were making rapid progress in identifying and fighting the virus, suggesting that a viable vaccine would soon be available and extolled the miracles that their medicine was producing. He had already greatly increased government spending for the National Institutes of Health after Rock Hudson died and would now double it between 1987 and 1989.[4]

Although some progress was noticeable in the treatment of "opportunistic infections" that afflicted AIDS patients, the disease could still not be treated with the largely ineffective AZT. By this time the bulk of AIDS research funding had shifted to the National Institute for Allergy and Infectious Diseases; the NIAID was better equipped than the National Cancer Institute to create a national system for coordinating, funding, and directing research to find a treatment against AIDS.

Tony Fauci, a determined immunologist, became the public face of the National Institutes of Health. He was not afraid of stepping into the limelight to engage with gay activist organizations. One of the loudest among them, ACT UP, was founded in 1987 by charismatic playwright Larry Kramer. The AIDS Coalition to Unleash Power had but one objective: to get drugs to those who needed it most. In order to shake up the establishment, it angrily took to the streets. The organization became known for its public disturbance, attention-drawing antics and media stunts such as the creation of traffic jams or the disruption of the communion mass at Saint Patrick's Cathedral. Other gay associations were more subtle and felt that cooperation and persuasion would be more productive. But Kramer was out to attack everybody including the pharmaceutical companies. ACT UP started by placing pressure on Burroughs Wellcome to lower the price of

AZT from $10,000 to $6,400 a year. Following their first successes, they would demand a seat at the table in many corporate boardrooms and a voice in the FDA decision-making. The high levels of media coverage dramatically raised public awareness of the ongoing tragedy.

AIDS had also become the focus of the nucleosides network, Richard Walker and Erik De Clercq had created. Their third NATO conference was held in May 1987 in Il Ciocco, a beautiful resort in the heart of Tuscany overlooking the medieval town of Barga, once Michelangelo's marble workshop.

It came as no surprise that the keynote speaker would be the fifty year-old Robert Gallo, now at the peak of his fame. He had restored his good name thanks to the joint announcement of President Reagan and Prime Minister Chirac a few weeks earlier. The settlement with his rival Luc Montagnier felt "like a piece of lead came off my shoulder," Gallo told reporters.[5] Not long before, he had been the guest of honor at Leuven University where he was awarded an honorary doctorate for his discovery of the human retrovirus that caused leukemia. His detailed descriptions of the behavior of a human retrovirus were crucially important in aiding the science community to tackle HIV.[6]

George Galasso was another stirrer of commotion, albeit on a different level: his talk was a pressing call to arms, to find the drugs to combat HIV immediately. Galasso was one of the leading figures at the NIH. He had himself worked with interferon and was an early supporter of recombinant interferon. He supported many clinical studies through NIH funding. Because of the very effective way in which he assisted drug research and development, he was much appreciated in the nucleosides community. When he spoke at the conference in Il Ciocco, the audience listened closely and took his advice to heart. Gallasso expressed what was on everybody's mind:

> AIDS has helped antiviral drug development. We have made giant advances thanks to AIDS. The prevailing skepticism of the past has given way to optimism and determination. This determination is partially due to recent successes in the field, but more likely the results from the AIDS epidemic. We did not realize the severity of AIDS, the causative agent was just being identified as a virus. It is most unfortunate that a

> disease such as AIDS proved to be the needed stimulus to advance antiviral research. We now know more about HIV than any other virus or cell thanks to the urgency of AIDS.

He encouraged the scientists to concentrate on one goal: to identify essential components of the virus specific to HIV and develop methods of blocking them.[7]

It was the first time John Martin was able to attend a NATO-ASI. As a representative of Bristol-Myers, he encountered two of his "drugs-in-the-making" in the room: Bill Prusoff of the Yale University and the Holý-De Clercq tandem. Prusoff had clinched the victory in the race to file the patent for d4T. This aided Bristol-Myers's neighborly relations. At a 20 minute distance from each other, Wallingford and Yale University, New Haven were geographically very close.

But John Martin was not impressed by the proceedings of the ten day conference. He shared the scepticism of his friend, Richard Whitley, regarding the strict rules of the NATO-ASI's. Together with George Galasso from NIH they fomented a quiet revolt and laid the groundwork for a new nucleosides network. It brought together the same mix of people, but more frequently, in annual conferences and without a limitation on the number of American participants. So as not to compete with NATO, they simply convinced De Clercq to join their new venture and ignored Walker's objections. Thus the third NATO ASI on nucleosides gave birth to a much larger organization: the International Society for Antiviral Research (ISAR). With its annual ICAR conferences, it has become a thriving community of scientists.

Not all participants in Il Ciocco were involved in the projects that were taking shape behind the scenes, they very much enjoyed the lighter moments. An outdoor swimming pool was quickly becoming the main attraction. After all, dainty ladies swimming topless was not something these scientists were used to seeing every day. They also looked forward to one of the highlights of the conference, the excursion to the coast. It was a leisurely drive through the magical Italian landscapes and towns disgorging history, artisanship and colorful markets.

But when they reached the beaches of Marina di Pietrasanta, the group was shaken by a drama that unfolded in front of their eyes. The Japanese fellow of the Rega Institute, Takashi Sakuma, was taken by an undertow and about to drown in the Thyrrenean Sea. Rudi Pauwels jumped in the water and dragged the almost lifeless body to shore and resuscitated him with chest compressions until lifeguards brought him to the hospital. It was not the first time Rudi Pauwels saved someone's life. He had saved a technician who had fallen out of a kayak on a turbulent river in the South of Belgium. The incident solidified Pauwels "can do" reputation forever. His homecoming to Leuven was nothing short of triumphant.

Holý's compound is active against HIV

Nothing daunted or deterred Rudi Pauwels. His energy was unwavering, his output prodigious. Now that Janssen's financial support was kicking in, he became even more zealous. The two man-team he formed with his Japanese colleague and friend, Masanori Baba, attracted other talents as well. A young immunologist, Dominique Schols, liked the idea of becoming a "Janssen fellow." Coopting him would not only result in an inflow of new skills for the virology department, but also meant acquiring equipment that was previously reserved for the immunology department only: the FACS machine, a top of the line fluorescent automated cell sorter, which Schols adapted for their HIV research.

Erik De Clercq could not be more pleased. He felt his dream team was taking shape; Dominique Schols became involved with two HIV laboratories and liaised between them. Both groups always coalesced around De Clercq's treasure chest—the room where he kept his vast and ever expanding collection of compounds neatly stored in refrigerators and from where he issued instructions to each of his technicians and assistants. The enthusiasm and energy generated by the new team was infectious, and attracted an even wider circle of young scientists. Together, they worked long hours, sometimes through the night.

On one of Rudi's forays into Erik's treasure chest, he found the compounds from Prague and started to assay PMEA. Its antiviral activity had already been known, but Rudi discovered it could work against HIV as well! Years later, the compound also became a hepatitis B treatment.[8] Erik De Clercq found another compound from the Holý collection, HPMPC, to be a strong antiviral, albeit not against HIV. A decade later, the FDA approved a drug based on that compound for retinitis in AIDS patients.[9]

Almost simultaneously a package of compounds from the Showa University in Tokyo arrived, courtesy of the Japanese co-workers in Richard Walker's lab. The Japanese compounds were named HEPT and resembled acyclovir; the next logical step was to test whether they could be effective against the herpes viruses. But De Clercq's intuition told him that they would not, which proved indeed to be the case. Masanori Baba then looked at the Japanese compounds in Rudi's HIV lab, repeatedly testing them. They concluded that HEPT belonged to a new class of antivirals with anti-HIV activity. Similar compounds were discovered at Merck as well and were given the cumbersome name, non-nucleoside reverse transcriptase inhibitors (NNRTI). This new class of antivirals used a different method to thwart HIV's invasion of human DNA than the Acyclic Nucleoside Phosphonates. Soon, other non-nucleosides were also discovered, but this time from the Janssen library.

All these exciting discoveries raised the curtain for a new generation of drugs. They had unmistakable influenced and quickened the pace of the triangular negotiations with Bristol-Myers and the Czechoslovak Academy of Sciences. Everything was readied for the concluding session in Prague: Erik De Clercq as the representative of the Rega Foundation and Julius Vida representing Bristol-Myers were the two foreigners confronting a gaggle of more than twenty Czech lawyers, Communist party bosses as well as the leadership of IOCB and members of *Inventia*, the Office for the Protection of Intellectual Property; all were in a combative mood.

They were determined not to repeat the bungling that ensued from the licensing of Otto Wichterle's soft lens to an American company. The deal was poorly negotiated by the Academy of Sciences, at that time still under the leadership of František Šorm. The most comprehensive technology transfer from East to West in the sixties had developed over the years into a tremendous commercial success. Both the Academy in Prague and the inventor had remained largely unrewarded.

But before tackling Bristol-Myers, the Czech lawyers wanted to review the patents that De Clercq and Holý shared. They proposed that the chemist receive a higher reward than the virologist. After all it is the chemist who synthesizes the compound. Who is the creator, the inventor?

Tony Holý veered up as if stung by a hornet's nest. He was genuinely angry. He would not tolerate that either the chemist or the biologist take priority over the other. "Both are due equal shares!" he emphatically said and the debate was closed.

Then the lawyers confronted Julius Vida. Royalties were due not only on the basis of drugs developed but also on the basis of all prodrugs that would be derived from the compounds. If Bristol-Myers would not develop these compounds, the whole class of Holý's compounds had to be returned including all results and calculations of all the tests performed within the company. In order to close the deal, Bristol-Myers had to finance the trips of two communist party executives to the United States.

Vida found this rather amusing but did not object; Czechoslovakia under the Husàk regime remained the most hardline country in the socialist sphere, defying Moscow's recent turnabout. Since Mikhail Gorbachev had come to power in 1985, the Soviet Union had been steadily progressing towards restructuring (*Perestroika*) and opening (*Glasnost*) communist society. Gorbachev adopted a laissez-faire approach in dealing with the rest of the Warsaw Pact countries. The changing tide of Soviet priorities placed the Husàk party in an awkward position, but the regime shrewdly ensured that Czechoslovakia remained hardline without having to resort to all-out political terror. In the Academy of Sciences, the winds of change were starting to blow and Julius Vida was happy to oblige.

The license to Bristol-Myers meant that the Holý compounds were now going to be intensely scrutinized. Cooperation between the pharma giant and academic institutions became redundant. Holý and De Clercq had to wait more than two years before Bristol-Myers informed them about the future of these compounds; it was a nail-biting time.

When Nobel prizes were announced in the fall of 1988, the field of Nucleosides research finally earned its share in the limelight. A Nobel Prize was awarded to Gertrude Elion and her boss George Hitchings at Burroughs Wellcome for the innovative ways in which they had developed a series of drugs. The Nucleosides community believed that her Nobel Prize was triggered by her discovery of the mechanism of the antiviral activity in acyclovir.[10] Although much attention was not paid to Elion before, the Nobel Prize changed everything. She was now the figurehead of Nucleosides research. Before she retired in 1983, Elion was the head of the Department of Experimental Therapy and had a hard time convincing her colleagues at Burroughs Wellcome to actually develop the drug.[11] Richard Whitley who conducted clinical trials with acyclovir was her closest ally. De Clercq and his friends at Rega helped her drug to become even more popular. They found that the amino acyl esters of acyclovir had better aqueous solubility and could be developed as a better alternative. They received a worldwide-minus-the US-patent for their findings. De Clercq was able to license these to Wellcome, the British parent company of Burroughs Wellcome. Their discovery was superseded in the U.S. by the patent for one of the amino acyl esters, the valine ester, that facilitated oral absorption. In De Clercq's eyes, these were merely "me-too" drugs and he never touted Rega's contribution to the acyclovir story very much. Patent rights were nevertheless granted from 1995 to 2002 before valacyclovir turned generic.

Chapter XII
Finding the best therapy: the one-a-day-pill

The creative act is not an act of creation in the sense of the Old Testament. It does not create something out of nothing; it uncovers, selects, re-shuffles, combines, synthesizes already existing facts, ideas, faculties, skills. The more familiar the parts, the more striking the new whole.
— Arthur Koestler

A new start-up: Gilead Sciences

The biotech gold rush on Wall Street had been unleashed. Companies like Biogen and Genentech had captured the imagination of investment bankers even before a single product was made. Recombinant DNA technologies required to produce interferon synthetically were adapted for use in other proteins. Anything seemed possible.

The excitement had captivated Michael Riordan, a young student at the Johns Hopkins Medical School. Born and raised in Kansas, son of a physician and a mother who wrote textbooks about breastfeeding for medical professionals, he was immensely curious about nucleic acids research. He gravitated very naturally to the world of interferons and gene expression. His favorite place to research was in the Johns Hopkins laboratory of Paula Pitha, a Czech virologist who had fled communism in the late 1960's.[1]

When Riordan graduated from Johns Hopkins with high honors, he toyed with ways of combining theoretical science from the academic world with the product development of the pharmaceutical industry. His next step, an MBA at Harvard, launched him on a different path, the world of venture capital. He was hired by Menlo Ventures in Silicon Valley. The hub of technology and innovation brought him closer to where the action was. He spent a whole year traveling the country, visiting pharma companies and academic institutions, and learning who the players were in the field of DNA chemistry.

In 1987, with $2 million of seed capital from his friends at the venture capital firm, he made the jump to start his own company. He named it Gilead Sciences, after the ancient site of a willow tree that produced a curative balm. The company began as a small lab outside San Francisco with just six employees. Very early on Michael Riordan managed to coax Gordon Moore, the co-founder of Intel, to join his business advisory board.[2]

Soon, it was time to install a directors' board. Undaunted, he went straight to the top, and chased the former Secretary of Defense, Donald Rumsfeld. "How many people work in your company?" Rumsfeld asked. "Just six including the founder" answered the twenty-nine year old Riordan. Rumsfeld who had just stepped down as CEO of Searle was impressed by his youth and determination and happy to impart his pharma experience to a start-up in Silicon Valley.

At the start, none of Gilead's experimental drugs worked outside the laboratory, but the biotech craze was in full swing and his ideas caught the interest of the venture capital world. Riordan remained focused, unperturbed and continued to entice more investors for his search to find drugs to control disease-causing genes.[3] One of them was Benno Schmidt, a partner of J.H. Whitney & Company. He had been an influential powerbroker in New York City when President Nixon appointed him to the chairmanship of the President's Cancer Panel, which initiated the federal government's "War on Cancer." Benno Schmidt pushed his firm to invest in biotechnology ventures. As a leader in both the private and public sector, he was considered the "senior gatekeeper of biomedical innovation in the United States."[4] Once Benno Schmidt was on board, Riordan was able to attract capital infusions from Venrock, the Rockefeller investment firm and Glaxo, the pharmaceutical company.

Next, Riordan felt they needed to add a person of stature in Europe to enhance the board of directors' international prominence. Rumsfeld thought of his Belgian friend, Stevie Davignon, whom he had met during his time at NATO and later became a Vice President of the European Commission.[5] Rumsfeld attracted a few other big names and later also George Schultz, the Secretary of State under President Reagan.

Capital was no longer a problem, but now Riordan desperately needed a group of first rate scientists. 1990 became a "golden year" when the main players of his team would fall into place. Riordan plucked Norbert Bischofberger away from Genentech, and scooped up John Milligan right after he finished his postdoc at UCSF. He spent many months, but eventually pryed Bill Lee away from Syntex.

He was still on the hunt for a Head of Research, the thirtieth employee. Riordan was looking for someone who could bring Gilead to the next level, and support the development of innovative drugs. He focused on a charismatic person, steeped in science and with a passion for entrepeneurship. He interviewed all the potential hires personally over dinner paired with fine wines. After many dinners and plenty of wine, he still could not find the right person. At least not until a headhunter drew his attention to the newly merged Bristol-Myers and Squibb and raised the possibility of hiring the director of infective chemistry, John Martin.

One of Riordan's scientific advisors, Richard Whitley, made it happen. He prepared the terrain and sweettalked John Martin before he would sit down with Riordan over dinner in the Smith & Wollensky restaurant, a popular steakhouse in Manhattan. Perhaps not the most ideal venue for a person like Riordan who had only just become a vegetarian.

However, three hours later everything clicked. They had concocted a plan for John Martin's smooth transition from Bristol-Myers to Gilead. Martin pointed Riordan to his co-workers, Swami Swaminathan and Mick Hitchcock, both at Bristol-Myers. Both were hired that same year.

Another part of Riordan's strategy was to visit Erik De Clercq and Tony Holý. Things had to be kept confidential. When Riordan arrived in Prague and Leuven, his stopovers seemed like simple courtesy visits at the time. In fact, they turned out to be reconnaissance for future collaboration.

Just a few months later, De Clercq and Holý felt the ground shift beneath them. They were invited to Wallingford where they were informed that Bristol-Myers Squibb no longer wanted to develop the Acyclic Nucleosides Phosphonates. The news hit them like a ton of bricks! One week after the announcement by Bristol-Myers-Squibb on 20 May 1991, John Martin was on the phone with Holý and De Clercq trying to convince them to transfer their license to Gilead. Allotting their license to a start-up in California seemed like a particularly risky affair—most of these new

Silicon Valley companies went belly-up and could not be trusted. But something told them this adventure could be different. Their positive experience working with John Martin convinced them to consider the offer.

A few weeks later, they agreed to meet in Paris in a restaurant not far from the Tuileries on July 2, 1991. The negotiations were very straightforward, they would stick to the exact same licensing agreement they had concluded a few years earlier with Bristol-Myers.

Riordan, Martin, De Clercq and Holý sealed the deal by signing on a table napkin. The formalities of officially transferring the license followed soon after.

In Wallingford, Julius Vida was distraught over the way Squibb handled the merger and imposed their will on Bristol-Myers. He bemoaned their killing of the goose that laid the golden egg. It was with a heavy heart that he saw the phosphonates leave the company. Nevertheless, Julius Vida was relieved they could be developed under John Martin's stewardship. Vida played a crucial role in facilitating the transfer of the license and making sure it involved every test, every calculation, and that every single piece of information acquired under Bristol-Myers's watch was handed over.

Michael Riordan did not want anything to go awry. He traveled to Wallingford to take hold of the dossiers in person and send them by Express mail to California. Only at that moment did he feel assured of the company's new beginning. Gilead was off to a flying start! Though it had not yet developed any products or posted any profit, the thirty-odd employee company basked in the confidence and promise of the newly acquired intellectual property: the acyclic nucleoside phosphonates stemming from the collaboration between the Rega Institute and the IOCB in Prague.

At the end of that same year, 1991, Riordan began the filing process for an initial public offering with the Securities and Exchange Commission. It was completed three months later, heavily oversubscribed and signaled the public market's unofficial stamp of approval.[6]

The birth of Cidofovir, Tenofovir and Adefovir

Acquiring the license from Bristol-Myers meant John Martin and his colleagues could simply pick up from where they left off in Connecticut. It saved them years of research and bolstered their confidence immensely.

But when Martin was introduced to Paul Janssen at a Gordon conference in March 1992, the famous Belgian drug maker poured cold water all over it. He tempered his enthusiasm about the new company Gilead and gave him only one in a hundred chances to succeed.[7]

Riordan and Martin used their resources in a savvy way. They did not spend their time building up their internal capacities but made judicial use of outsourcing. In the early years, this philosophy involved intensive cooperation with their academic partners, IOCB in Prague and the Rega Institute in Leuven. The young company needed all the optimism and energy it could muster in order to withstand the roller coaster that was to follow.

There was no abating the AIDS epidemic. By 1991, more than 100,000 Americans had died from the disease, nearly twice as many as had perished in the Vietnam War. People were dying more from the "opportunistic" infections rather than the disease itself. Gilead's first priority was to tackle HPMPC in the hopes that it could be effective against the cytomegalovirus (CMV), a virus of the herpes family. CMV did not cause disease in healthy people but was life threatening to the immunosuppressed. It caused blindness, pneumonia, severe diarrhoea and encephalitis. John Martin knew the disease well since he had synthesized gancyclovir while at Syntex.[8]

Just as HPMPC was on its way to becoming cidofovir, it almost tanked the company: cidofovir was causing cancer in rats! It was one of those all hands on deck situations. Erik De Clercq had to travel urgently to Foster City and finetune the dosage to reduce its toxicity. He was greatly helped by his co-workers in Leuven.[9] John Martin had to use all his persuasive powers to keep the FDA engaged.

At the Rega Institute, more compounds kept arriving. Non-nucleosides from Janssen were screened and scrutinized in Rudi Pauwels's lab.[10] The nucleotides from Prague were screened and analyzed in Jan Balzarini's lab. One day, a new compound from Prague, named PMPA, was thrown in the mix. Holý always gave De Clercq the privilege of naming his compounds, since he usually oversaw all the operations and documented them in publications. When the article about PMPA appeared in 1993, it was already obvious that the compound had a high capacity to fight HIV—100 times greater than AZT—and a very low toxicity.

With an amendment to the orginal license, PMPA was added to Gilead's basket of more than 500 Holý-De Clercq compounds. It became the focus of the PMEA-team led by Mick Hitchcock. The former co-worker of John Martin's at Bristol-Myers had a special intuition for testing and screening, dating back to his days with interferon and later with d4T. PMPA arrived at a critical time in Gilead when it evolved into tenofovir. PMEA was not the anti-HIV agent that it was extolled to be and its toxicity caused concern at the FDA. Before it could endanger the reputation of the company, PMEA was quietly moved to the backburner and re-examined several years later. At a much lower dosage, the compound evolved into adefovir and was ideal for treating the hepatitis B virus. In 10% of the cases it could actually kill the virus.

In the midst of all the activity of readying Gilead's cidofovir for its approval by the FDA, news of a new anti-HIV treatment shook the scientific community. Three drugs appeared almost simultaneously on the market in 1994, capable of attacking the human immune deficiency virus from another angle, through its protease enzyme.[11]

After AZT which inhibits the reverse transcriptase enzyme of HIV, these new drugs were another reprieve for people with AIDS. Though short-lived, it kept the hope alive until a miracle breakthrough was announced in 1996 during the International AIDS conference in Vancouver. HAART, the highly active anti-retroviral treatment, was the result of hundreds of researchers toiling in university and pharmaceutical labs. The groundbreaking combination therapy was spearheaded by David Ho, the Director of the Aaron Diamond AIDS Research Center in New York.[12] His insights into the reproduction of the virus turned the tide of AIDS.

Until then it was thought that the virus remained in a long, almost dormant, state before it attacked the body's immune system in full force. Instead, David Ho found that the virus started replicating and mutating itself furiously as soon as it had entered the body. HAART's success was due to its ability to block two of the virus's crucial enzymes through the combination of nucleoside and non-nucleoside reverse transcriptase as well as protease inhibitors.

It involved a daunting regimen—requiring some 30 pills per day to be taken at specific times; some with milk and others without, some before a meal and others after. The results were immediate, people who had prepared themselves to die were suddenly faced with a new lease on life. The deathtoll that year dropped immediately from 50,000 to 20,000 deaths. Even though it was very hard to adhere to, HAART paved the way for companies like Gilead to design a drug that was both effective and easy to comply with.

This became possible after the first Gilead tests with monkeys had proven 100% successful in 1995. The company gained stature once cidofovir was approved in 1996 and introduced to the market under the brand *Vistide*. The success of the company reflected itself in Michael Riordan's ability to raise close to $500 million after several more rounds of public offerings. With John Martin on board, Riordan felt the company was now in safe hands and proposed he be named the CEO of Gilead. The time had come for Riordan to pursue other dreams. By the time he left, he had hired 250 employees.

Gilead ramped up its antiretroviral profile and prepared for a spectacular entry on the market of *Viread* in 2001. It was the first one-a-day pill to treat HIV infections. At its core, *Viread* was based on the acyclic nucleoside phosphonates or acyclic nucleotides that were invented in Prague, analyzed in Leuven and developed in Foster City. *Tenofovir* and its prodrug TDF (Tenofovir Disoproxyl Fumarate) was the product of Tony Holý's genius, the perspicacity embedded in Erik De Clercq's group, and the driving force of John Martin in motivating his team to develop the compound.

As its next step, Gilead purchased Triangle Pharmaceuticals in 2003 and introduced *Emtriva* (FTC), followed by another miracle pill, *Truvada* wich combined TDF and FTC in 2004 and *Atripla* which combined three drugs (TDF/FTC/EFV standing for Efavirenz) into a single pill in 2006.[13]

Tenofovir would remain the cornerstone of all other anti-HIV pills made by Gilead. Even the second generation anti-HIV pills introduced in 2017, based on the prodrug Tenofovir Alafenamide Fumarate, the so-called TAF generation developed by William (Bill) Lee and Thomas Cihlar, contains *tenofovir* albeit at a much lower dosage. To this day, Gilead's drugs remain the gold standard for HIV treatment not only in the developing world but as of recent also in the developing world. In 2012, the FDA recognized *Truvada* as the ideal pill for prevention. If enough people had access to it, *Truvada* could put an end to the epidemic.

Epilogue
Of scientists and crusaders

What one needs in life are the pessimism of intelligence and the optimism of will.
— André de Staercke, former Belgian ambassador to NATO

Twenty years into the AIDS epidemic, little had been accomplished to thwart the disease in developing countries. With the exception of some minor milestones in the field of bloodscreening and educational efforts, HIV continued to spread like wildfire around the world. The situation was particularly disastrous in sub-Saharan Africa.

The tepid international commitment in the 1980's and 90's was partly due to the fact that only very ineffective drugs were available. The problems were compounded by the absence of political will, denial by leaders in the most affected countries and lack of funding. In the United Nations family very few multilateral organisations had been monitoring the crisis. Six of them finally decided to coordinate their actions in the field of HIV/AIDS and created UN-AIDS in 1995. Belgian scientist, Peter Piot, became the head of the coordinating agency, he raised awareness slowly but surely.[1]

The tipping point occurred in January 2000 when U.S. ambassador to the UN, Richard Holbrooke, persuaded his colleagues to convene a meeting of the Security Council concentrating on Africa. The meeting woke people up. The epidemic in Africa, ground zero of HIV/AIDS constituted a new type of security challenge. More than 11 million AIDS orphans could easily become weaponized as child soldiers. The diminishing demographics and drastic reduction in life expectancies were upsetting the political and economic stability of almost all affected countries.

The formal reason for putting "AIDS in Africa" on the agenda of the Security Council was the danger that peacekeeping operations posed in spreading the virus.[2] The Council adopted a resolution in that regard, but the meeting had a much larger outcome. From then on, HIV/AIDS would be dealt with at the highest level of government. It spawned a series of regional initiatives, most notably the summit on AIDS in Africa convened by the Organization of the African Union. As Peter Piot recounts in his book:

> One head of state after the other broke the silence on AIDS in their country, collectively the continent acknowledged at last, that it had an AIDS problem.

Sensing the support from the international community, the heads of state were now committed to tackling the epidemic. Shortly before the summit, UNAIDS and WHO had negotiated major price reductions for antiretrovirals with the pharmaceutical industry, and the first Indian generic antiretrovirals arrived on the African market.[3]

Funding, however, was still a major problem. The UN called for the establishment of a special fund for AIDS in 2001, which became the "Global Fund to fight AIDS, Tuberculosis and Malaria" a year later. Donors, led by the United States felt the UN was too slow and inefficient to manage an emergency fund and insisted the Global Fund would be established as a public-private partnership. The United States became its first supporter with a contribution of $200 million and pledged that it would match every other contribution.

The Global Fund has become a major game changer and so has the groundbreaking President's Emergency Plan for AIDS Relief (PEPFAR). President George Bush Jr. took everybody by surprise in his State of the Union address on January 29, 2003:

> Today on the continent of Africa nearly 30 million people have the AIDS virus, including three million children under the age of 15. There are whole countries in Africa where more than one-third of the population carries the infection. More than four million require immediate drug treatment. Yet, across that continent only 50,000 AIDS victims are receiving the medicine they need. But the cost of antretroviral drugs has dropped drastically which places a tremendous possibility within our grasp. Seldom has history offered a greater opportunity to do so much for so many.

His speech was the parting shot for America's involvement not seen since the Marshall Plan. At stake was nothing less but the saving of a generation in Africa.[4] The U.S. Secretary of Health and Human Services, Tommy Thompson, was one of those charismatic leaders who mobilized both public officials and private business to work on PEPFAR. He took them along on his missions to Africa and showed them how their contributions could make the difference. He impressed on them with dramatic effect that: "It's like a war, only this war is taking 3 million lives a year!".

It was on one of those trips, in the company of Tommy Thompson, that John Martin became alerted to AIDS in Africa. It touched him to the core. Upon his return to Foster City, he decided Gilead would ship its drugs, at discounted prices, directly from its manufacturing facilities in the U.S., Canada, and Europe to public and private organizations located in the most affected countries. At the same time, John Martin convinced the University of Leuven and the Academy of Sciences in Prague to relinquish their royalty rights for most of the developing countries.

The first access programs, however, were doomed to failure. Gilead had been emulating other companies, but did not make any headway: it looked more like PR. John Martin realized that a purely philanthropic effort was not sustainable. The company had underestimated the unique challenges facing drug-delivery in poorer countries. Martin also recognized that on its own, the company did not have sufficient capacity to meet global needs for HIV-treatment. He could have abandoned Gilead's laudable efforts there and then.

But giving up is not in John Martin's nature. Trial and error rekindled the same spirit that had energized the company since its very beginnings: collaborate and compete. Only through partnerships and collaborations would it be possible to increase drug access.

In 2006, Gilead entered into licensing agreements with Indian manufacturers, granting them rights to produce and sell high-quality, low-cost generic versions of its medicines. Currently 15 Indian manufacturers, one South African and two Chinese companies hold licenses. It became a major success.

All licensees hold either World Health Organizations pre-qualifications or FDA tentative approvals and the vast majority of Gilead's HIV drugs in developing countries—over 98 percent—are now generics produced by licensees. Partners receive a full technology transfer of the Gilead manufacturing process, enabling them to quickly scale up production. Gilead became the first—and to this day the only pharmaceutical company—to sign an agreement with the Medicines Patent Pool (MPP), an international organization that expands access to medicines through the sharing of drug patents.

As a next step, Gilead pursued a market-based strategy to provide access in developing countries to every patient who needs treatment. As long as Gilead's products were more expensive, the WHO guidelines favored d4T and it remained the drug of choice for a long time. This was especially painful since d4T was associated with greater toxicities, a higher level of patient discontinuation, hospitalisations and deaths due to AIDS. Fortunately things eventually changed. As of November 2016, more than 10 million people in low-and middle-income countries are now receiving Gilead HIV drugs, better suited for leading longer and healthier lives.

John Martin's "collaborative commercialization" strategy, finetuned by Gregg Alton, was what turned the tables around. The extension of non-exclusive licenses to multiple manufacturers promoted competition to produce large volumes of high-quality drugs and lowered the prices dramatically. It was this strategy, which enabled the sustained flow of drugs in the most efficient manner, that saved millions of lives. The World Health Organization estimates that new infections fell by 35% by 2015 and AIDS related deaths fell by 28%. The WHO aims to put an end to the epidemic by 2030.[5]

This ambitious goal is based on the fact that a new generation of Gilead's drugs is becoming available to low and middle income countries. The drugs still get their potency from the Tenofovir component albeit with a lower dosage.

Gilead is still a relatively small company with only 8.000 employees worldwide and very few brick and mortar assets in third world countries. Notwithstanding its size, it has given humankind the know- how to develop the acyclic nucleoside phosphonates, otherwise called acyclic nucleotides. These compounds were created by the unique collaboration between Tony Holý and Erik De Clercq. But it was John Martin who recognized the extraordinary capability of acyclic nucleotides and led a remarkable team of scientists who developed them into highly effective drugs. This breakthrough achievement has enabled more than 90 % of HIV patients in the western world and a growing majority of patients in developing countries to lead normal lives. Thanks to the powerful desire to succeed by John C. Martin and his colleagues, more than eight million lives have already been saved. The new generation of drugs that they developed will continue saving many more lives around the world and promises to serve as the world's greatest hope in bringing an end to the HIV epidemic.

Tony Holý with self-made canisters around 1970. His laboratory was very sparse and many reagents were homemade (Photo courtesy of Dana Holà archives).

Tony Holý and his technicians in his laboratory around 1980. The tall lady in the background, Bela Novakova, taught him many skills and remained very devoted to him all along his career (Photo courtesy of Dana Holà archives).

Erik De Clercq, teaching biochemistry on the campus of KU Leuven in Kortrijk, around 1980 (Photo courtesy of Erik De Clercq archives).

Symposium of a group of chemists in Kyoto around 1982. Nobel Prize laureate Gobind Khorana seated in second row, third from left. Last row: Richard (Dick) Walker (fifth from the right), Tony Holý (third from right) and Erik De Clercq (second from right) (Photo courtesy of Erik De Clercq archives).

Rudi Pauwels, who started the first AIDS laboratory at the Rega Institute in 1986. His system was copied in laboratories all over the world for more than twenty years. He was still a student at the pharmaceutical faculty at that time and obtained his Ph. D. in 1990 with the highest honors (Photo courtesy of Erik De Clercq archives).

Two Japanese doctoral fellows at the Rega Institute: Masanori Baba on the left discovered antiviral activity in Piet Herdewijn's compounds and was instrumental in other discoveries together with Rudi Pauwels. Takashi Sakuma, on the right, discovered the anti-VZV activity in the very first acyclic nucleoside phosphonates (Photo courtesy of Erik De Clercq archives).

In a Brussels restaurant around 1992. From left to right: John C. Martin, then Head of Research at Gilead Sciences, Etienne (Stevie) Davignon, Member of the Board of Directors of Gilead Sciences, Norbert Bischofberger, then Head of Development at Gilead Sciences, Michael Riordan, founder and first CEO of Gilead Sciences, and Erik De Clercq, Head of the Rega Foundation and Head of the Institute (Photo courtesy of Erik De Clercq archives).

Photo taken at Gilead Sciences in 1992. From left to right: Erik De Clercq, John C. Martin, Tony Holý (Photo courtesy of Gilead Sciences).

Gilead Sciences maintained close contacts with the Rega Institute all through the 1990's into the new Millennium. Here John C.Martin on one of his regular visits to Leuven around 1995. From left to right: Johan Neyts, then a postdoctoral fellow working on the potential of Cidofovir analogues to fight smallpox. Robert Snoeck was the clinician for most testings. He investigated the potential of cidofovir on viruses including the CMV virus. Lieve Naessens worked with Gilead Sciences on the development of the prodrug of Tenofovir. Jan Desmyter, then the head of clinical virology in the university hospital, was a vital supporter of the AIDS laboratories in the virology department of the Rega Institute. John C. Martin, then CEO of Gilead Sciences and Norbert Bischofberger, then the Head of Research at Gilead Sciences, Erik De Clercq then head of the Rega Foundation and head of the Institute, Graciela Andrei, who was instrumental in the discovery of the antiviral action of Cidovofir in papilloma viruses (Photo courtesy of Erik De Clercq archives).

Launching of VISTIDE®
by Gilead Sciences on 27 June 1996

The launching of "Vistide" in 1996, the first commercial product of Gilead Sciences. The company had fewer than 200 employees at that time. Today Gilead is still a relatively small pharmaceutical company with 8,000 employees worldwide (Photo courtesy of Gilead Sciences).

第9回国際抗ウイルス学会議
Ninth International Conference on Antiviral Research
May.21.1996 Urabandai

Ninth International Conference on Antiviral Research (ICAR), May 21 1996, in Ura-bandai, Northern district of Fukoshima. First row (seating) from right to left: John C. Martin, Erik De Clercq, Mrs Chikako Shigeta, Shiro Shigeta, Earl Kern, Hugh Field, George Galasso, Rich Whitley. Towering above the whole scene: Raymond Schinazi and Naoki Yamamoto on third row far left (Photo courtesy of Erik De Clercq archives).

From left to right: Tony Holý at a meeting in Atlanta in 1997, Mick Hitchcock who tested many compounds at Bristol-Myers and later at Gilead to determine which compounds to develop; Piet Herdewijn, presently Head of Medicinal Chemistry at the Rega Institute who synthesized d4T and many other nucleosides that are now part of Gilead Sciences's library (Photo courtesy of Erik De Clercq archives).

Tenth International Conference on Antiviral Research (ICAR), April 1997, in Atlanta, Georgia. First row from left to right: George Galasso, Mrs. J. Galasso, Bill Prusoff, Earl Kern, Raymond Schinazi, John C. Martin, Nobel Prize laurate Gertrude (Trudy) Elion, Rich Whitley and Hugh Field (Photo courtesy of Erik of De Clercq archives).

Tony Holý receiving the State Medal of Merit from the hands of president Vaclav Havel in 2002 (Photo courtesy Dana Holà archives).

Tony Holý, revered as a great Czech scientist, points to a Czech invention, the sugar cube. Czech inventiveness was one of the themes to promote the Czech presidency of the Council of the European Union in 2009 on posters throughout the Czech Republic (Photo by Vladimir Kopal).

Erik De Clercq and Tony Holý receiving honorary doctorates on 4th June, 2009, at the University of Southern Bohemia. Professor and later rector, Libor Grubhoffer, in their midst; John C. Martin, then CEO of Gilead Sciences at the left, Zdenek Havlas who became Tony Holý's successor as director of the IOCB on the right (Photo courtesy of Dana Holà archives).

Erik De Clercq and Donald Rumsfeld in 2000. Rumsfeld's last meeting of the board of directors at Gilead Sciences prior to his nomination as Secretary of Defense (Photo courtesy of Gilead sciences).

Intimate ceremony in garden of the Belgian Ambassador in Washington honoring John C. Martin, former CEO of Gilead Sciences and Executive Chairman of the Board of Directors. From left to right: Bill Lee responsible for the second generation of Tenofovir-based drugs of Gilead Sciences, the so-called TAF drugs; Koen Debackere, General Administrator and Vice-rector of the KU Leuven; Dirk Daelemans of the Rega Institute; Thomas Cihlar of Gilead Sciences who developed the TAF- drugs with Bill Lee; Gregg Alton, responsible for Gilead Sciences's reach in countries outside the U.S.; Swami Swaminathan, the mathematics wizzard at Gilead Sciences; Nicolas Gouwy, Representative of Gilead Sciences in Belgium (Photo by Renilde Loeckx).

Notes

Introduction

1. As a junior diplomat in Bonn, in the eighties, I was in charge of scientific matters in our embassy. My lack of scientific training was obvious to the German Minister of Science and Technology who put me at ease and reassured me: "Wir sind Alle Juristen." While serving in the Belgian embassy in Paris in the nineties, French scientists encouraged me to become more vocal about Belgium's contributions to science. As a Consul-general in New York, I had the honor of hosting Fundraising dinners for Belgian companies involved in Microbicide trials in Africa.
2. Renée Fox is the author of *Experiment Perilous,* a classic study of medical research. Her book *In the Belgian Château* not only provides an excellent window into Belgian academic medicine but in the country itself.
3. John C. Martin *et al.* (2010) "Early nucleoside reverse transcriptase inhibitors for the treatment of HIV".
4. See Dorothy H. Crawford (2007) *Deadly Companions.*
5. Pasteur was a great friend of the English surgeon Joseph Lister and helped him to propagate antiseptic methods for physicians to apply when treating patients. See Paul De Kruif (1926 & 1996) *Microbe Hunters.*
6. Anton Van Leeuwenhoek, a Dutch shopkeeper from Delft who in the seventeenth century first uncovered a whole menagerie of "animalcules" as he called the little creatures he saw crawling under his lenses. It was not before the second half of the nineteenth century when some of the secrets of these creatures, which Louis Pasteur termed "microbes", were unlocked. See Paul De Kruif (1926 & 1996) *Microbe Hunters.*
7. Robert Koch, a physician in Berlin, dismissed Pasteur and became his fierce rival. It was as if animosity from the Franco-Prussian war of 1870 had spilled over into their labs. Koch worked on his experiments in a systematic, coldly logical way following his postulates which researchers still abide by to this day. He was hailed as the "father of the microbial theory of disease" for his proving that specific germs caused specific diseases. Koch identified the bacteria

causing anthrax, tuberculosis and cholera but was not able to find a remedy. See Paul De Kruif (1926 & 1996) *Microbe Hunters.*

8 While arm-to-arm inoculation or variolation with the virulent smallpox virus was practiced in India and China long before it was introduced in Europe at the beginning of the 18th century. Inoculation sensibly lowered case fatalities of smallpox but the intervention was fraught with danger. See Stefan Riedel (2005) "Edward Jenner (1749–1823) and the history of smallpox and vaccination" *Baylor University Medical Center Proceedings* 18(1):21–25.

9 See Stefan Riedel (2005), *ibid.*

10 See Paul A. Offit (2007) *Vaccinated.*

11 At the dawn of the twentieth century the Dutch microbiologist Martin Beijerinck had isolated the tobacco mosaic virus and found that the virus could only live and propagate in plants. He deduced from a pure chemical analysis that a virus, the smallest of all microbes, had to be a parasite. Knowledge gained many years later showed his vision came surprisingly close to the modern concept of a virus. See A.P. Waterson and Lise Wilkinson (1978) *An Introduction to the History of Virology.*

12 The Enders group's technique is still used to make viral vaccines today. See Paul A. Offit (2007) *Vaccinated.*

13 See John Booss and Marilyn J. August (2013) *To Catch a Virus.*

Chapter 1. Leuven: a hotbed for antiviral research

1 Piet De Somer's boss was Richard Bruynoghe, co-owner of a small pharmaceutical company, *Soprolac* which was purchased by a young industrialist, Jacques Lannoy.

2 See Alfons Billiau (2009b) "Penicilline in België".

3 Fleming's description of *penicillium notatum* is considered as one of the most important medical papers ever written. See also Sir Alexander Fleming's speech at the Nobel Banquet in Stockholm, December 10, 1945 about his findings in the St Mary School in London in 1928, retrieved from http://www.nobelprize.org/nobel_prizes/medicine/laureates/1945/fleming-speech.html.

4 The fact that all three major universities in Belgium: Leuven, Brussels and Liège, almost simultaneously offered an honorary doctorate to Alexander Fleming in 1945, underscored their desire to obtain the penicillin formula.

5 Lannoy's company was first named RIST (*Recherche et Industrie de Synthèses Thérapeutiques*) and later renamed. After the intervention of an angry Parisian doctor of the same name, docteur Rist, it became simply RIT as narrated by one of De Somer's assistants, Alfons Billiau who recorded much of Leuven's medical history.

6 In England at that time, penicillin was also made by a dairy company, GLAXO, making use of their old milk bottles. After 1967, RIT became part of Smith Kline and was the steppingstone of Smith Kline's expansion on the continent and its merger with GLAXO, See Alfons Billiau (2009b), "Penicilline in België".

7 See de Duve's (2004) speech "The first milligrams of Belgian penicillin" at the festivities for the 50th anniversary of the Rega Institute.

8 Professor Joseph Hoet had excellent contacts with Charles Best, the co-discoverer of the insulin hormone (1921) and the Director of Connaught Medical Research Laboratories in Toronto. See Alfons Billiau (2009b), "Penicilline in België".

9 Piet De Somer also witnessed a flourishing cooperation between university research and a profitable laboratory created with Charles Best's insulin royalties. See Alfons Billiau (2009), "Penicilline in België".

10 Hans Christian Gram, a Danish bacteriologist, found in 1884 that bacteria are divided into two groups. He used a stain and showed that some cells retained the stain and others did not. Those that did not retain the stain, Gram negative bacteria caused diseases like typhoid, tuberculosis and cholera; Gram positive bacteria retained the stain and were bacteria such as staphylococcus, the common cause of blood poisoning against which penicillin is effective.

11 The *Laboratoire de Bactériologie* was located in the Vital Decosterstraat in Leuven. See Hubert Vanderhaeghe in De Clercq (ed) (1987) *Frontiers in Microbiology*.

12 Waksman had emigrated to America in the waning days of the Russian empire and was welcomed in the Agricultural Department of the Rutgers University. He was well versed in soil bacteria that produced nitrogen available for crops and sold his findings to the brewery and food industries. George Merck, the owner of the pharmaceutical company bearing the same name asked him to look for substances from microorganisms that could treat people. Waksman avoided the topic at first in a similar way he had dodged other opportunities to search for drugs. In 1932 the American National Association Against Tuberculosis had asked Waksman to investigate why tubercle bacilli are

rapidly destroyed in the soil. He simply confirmed that this phenomenon was probably due to the activity of other predator microbes but did not investigate further. Was he afraid to work with the tuberculosis bacteria? His lab was indeed very poorly equipped as described by one of Waksman's biographers. See Peter Pringle (2012) *Experiment Eleven.*

13 After Hubert Vanderhaeghe had joined the team with his impressive chemistry skills, De Somer's group developed Griseomycin (in 1953), Virginiamycin (in 1954) and Lo-mycine named after the soil of Kessel-Lo, a village near Leuven. Virginiamycin was later used as a food additive for poultry and swine.

14 See Alfons Billiau (2009), "Penicilline in België".

15 The Rega Institute was at first composed of three sections. Bacteriology and medicinal chemistry concentrated on the discovery of new antibiotics and new vaccines. A third one was a laboratory entirely devoted to virus research.

16 The Belgian provinces in the 1700s were then part of the Austrian Habsburg empire and ruled over by governors from Vienna. Hendrik Rega (1690–1754) a disciple of The Enlightenment enjoyed their trust. He developed the library as a pillar of knowledge. Sciences were then part of the philosophy department and covered wide-ranging interests. See Robert Halleux *et al.* (1998) *Geschiedenis van de wetenschappen in België van de Oudheid tot 1815.*

17 Hendrik Rega lived in a palace in the Parijsstraat, 74, in Leuven and is often depicted wearing silk coats, lace cravats and wigs. Rega gave his hometown several buildings like the famous library as well as the anatomy theater, replacing the one that was once used by Vesalius during his studies in Leuven. The botanical garden with the most exotic plants known in his time was another one of his legacies. It all testified to his many talents as a "natural philosopher." See *The History of Medicine Topographical Data* at Himetop.wikidot.com.

18 Monique Lamy was the principal assistant who studied equipment and the culturing of the poliovirus in the Danish laboratory of Herdis von Magnus and the laboratory of Pierre Lépine in the Pasteur Institute. See Alfons Billiau (2011) "Piet de Somer, het Leuvense Rega Instituut en het Belgische Poliovaccin in 1956–57".

19 In the second half of the twentieth century ten diseases were brought under control thanks to vaccines: polio, measles, mumps, rubella, chickenpox, hepatitis A, hepatitis B, pneumococcus, meningococcus, and Haemophilus influenza type b. See Paul A. Offit (2007) *Vaccinated.*

20 Awareness of the paralytic poliovirus became more focused when president Franklin Roosevelt started his second term in the White House and American organizations had embarked on a ferocious crusade to combat the virus. Roosevelt himself had become a victim at the age of 39, but the epidemic proportions of the virus only became apparent decades later. See David Oshinsky (2005) *Polio: An American Story.*

21 It was likely that this virus had circulated at low levels in human populations for thousands of years. A famous Egyptian stele portraying a priest with a withered leg is testimony to this. The outbreaks only reached epidemic proportions as of the early 1900's when waves of the disease hit cities and rural areas with relatively high standards of living. See David Oshinsky (2005) *Polio: An American Story.*

22 Americans saw the germ threat everywhere. Their obsessions with hygiene are amusingly narrated in David Oshinsky (2005) *Polio: An American Story.*

23 See Dorothy H. Crawford (2007) *Deadly Companions.*

24 Karl Landsteiner successfully isolated the poliovirus in 1908. De Somer had always been fascinated by this Viennese scientist, See Alfons Billiau (2011) "Piet de Somer, het Leuvense Rega Instituut en het Belgische Poliovaccin in 1956–57".

25 Salk came from a poor Jewish immigrant background, but was able to attend schools for the gifted. He had worked with his mentor at New York University on experiments with a killed-influenza virus. He built on this experience to inactivate poliovirus with formaldehyde and develop a vaccine in the Medical School of Pittsburgh. See Paul A. Offit (2007) *Vaccinated.*

26 Since the poliovirus had been shown to be host-specific—only humans and monkeys were prone to the disease—it was thought it could only grow on human and monkey tissue culture. Enders and his group opened the way for the discoveries of Salk and Sabin thanks to their discovery of the ability of polioviruses to grow in cultures of various types of tissue. They grew the viruses in test tubes using safe cell cultures. John Enders and his assistants received the Nobel Prize in Medicine in 1954. See Paul A. Offit (2007) *Vaccinated.*

27 See David Oshinsky (2005) *Polio: An American Story.*

28 De Somer was aided by one of his first assistants, Monique Lamy (1930–2007). She had gained experience in culturing the polio virus with Danish virologist Herdis von Magnus (1912–1992) in Copenhagen and Pierre Lépine

(1901–1989) at the Institut Pasteur in Paris. See Alfons Billiau (2011) "Piet de Somer, het Leuvense Rega Instituut en het Belgische Poliovaccin in 1956–57".

29 Culturing the virus in cells of monkey kidneys was one of Salk's great innovations. He also pioneered a method for killing the virus without destroying its outer enzyme, see David Oshinsky (2005) *Polio: An American Story*.

30 See Alfons Billiau (2011) "Piet de Somer, het Leuvense Rega Instituut en het Belgische Poliovaccin in 1956–57".

31 De Somer's assistant, Abel Prinzie, was sent to Pittsburgh and worked with Jonas Salk, in Tokyo he befriended Albert Sabin. See Alfons Billiau (2011) "Piet de Somer, het Leuvense Rega Instituut en het Belgische Poliovaccin in 1956–57".

32 To increase the credibility of his vaccines, De Somer wanted to make sure that his testing methods on small chicks acquired the same status as the Danish and Swedish standardized tests on guinea pigs or monkeys. The Rega Institute happened to have an abundance of three week old chicks as they were needed to study the growth potency of new antibiotics. In March 1958, Piet De Somer travelled to Pittsburgh to enlist Salk's help. The Belgian testing method was introduced to the World Health Organization in Geneva and was formally recognized as equivalent to Swedish and Danish methods soon after. See Alfons Billiau (2011) "Piet de Somer, het Leuvense Rega Instituut en het Belgische Poliovaccin in 1956–57".

33 World EXPO in Brussels took place from April till end of October 1958; the Fair attracted more than fifty million visitors over a period of six months.

34 A technique that had been pioneered by Max Theiler when he developed a vaccine against yellow fever in the 1930s. He succeeded in weakening human viruses by growing them in cells from other species. His yellow fever vaccine in mouse embryos in the mid-1930s is still used today. His method has been used for making weakened viral vaccines against measles, mumps, rubella and chickenpox. He won the Nobel Prize in Medicine in 1951. See www.nobelprize.org.

35 Albert Sabin was born in 1906 in Bialystok Poland, then in Imperial Russia. His Jewish family had survived the pogroms but fled the Russian famine during the Bolshevik regime in 1921 and arrived in the US when he was fifteen.

36 As the world is coming closer to the complete eradication of polio, the World Health Organization has asked its member countries in 2000 to revert to the killed virus vaccine so as to prevent any propagation of the virus. "In the late

1990s, the oral vaccine came in disrepute because of increasing evidence that the attenuated viruses, by replicating in the intestines of vaccinated individuals, can revert to virulence and provoke outbreaks of paralytic polio." See Alfons Billiau (2014), "A Polio Vaccine for Belgium in 1956".

Chapter II. Behind the iron curtain

1 The pilot of the U2 spy-plane, Gary Powers was taken in captivity. Eisenhower refused a public apology and Khrushchev abandoned his attempts to co-operate with the U.S. and waited for the inauguration in 1961 of the new President, John F. Kennedy.
2 *The Pugwash conferences, a staff analysis*, Committee on the Judiciary, U.S. Senate, 1961.
3 Scientists from East and West were preparing much of the groundwork for the Limited Test Ban Treaty to be signed a few years later in Geneva.
4 The election of a Republican president did not improve the atmosphere at first. Dwight Eisenhower had just entered office and refused to grant a pardon to Julius and Ethel Rosenberg for providing the knowledge of an atomic bomb to the Soviets. The shock of the Rosenberg executions in June 1953 reverberated far beyond the United States. Eisenhower surprised friend and enemy alike, a few months later, with his "Atoms for Peace" speech before the United Nations General Assembly. He spoke of his deep anxiety about an escalating nuclear arms race and the need to warn American people.
5 See David Oshinsky (2005) *Polio: An American Story*.
6 The fact that DNA holds the key to genetic information in cells is a discovery made by Oswald Avery in 1944. Avery connected Gregor Mendel's findings with those of another 19th-century discovery by Friedrich Miescher, a twenty-four year-old Swiss student working in a laboratory in the German town of Tübingen. The young student had found a new group of biological substances in the nucleus of the cell. Until then, chemists had classified the principal distinctive substances in living beings into three categories: fats, sugars and proteins. Miescher discovered the fourth one and called it a "nuclein" because he found it in the nucleus of the cell. He later renamed it "nucleic acids." See Ulf Lagerkvist (2003) *DNA Pioneers and Their Legacy*.
7 The Russian invitation to come to Moscow was issued in 1958, the same year the first education exchange agreements between the USSR and the USA

entered into effect. In 1957, the Soviets shocked the United States by becoming the first nation to launch a satellite into orbit around the earth. *Sputnik*, as it was called, frightened many Americans, who believed that the Soviets would soon develop an entire new class of weapons that could be fired from space. The Soviets had also sent the immortal HeLA cells into space to explore how tissues respond to zero gravity. See Dawn Field and Neil Davies (2015) *Biocode*.

8 The first public announcement of what Crick and Watson had found was made before "The Solvay Conference" in Brussels in April 1953, a month after Stalin died. The paper "Molecular Structure of Nucleic Acids: A Structure For Deoxyribose Nucleic Acid" appeared in the April 25th issue of *Nature* (Watson and Crick 1953). Watson presented the structure of DNA in June 1953 at the Cold Spring Harbor meeting on Long Island, one of the birthplaces of modern biology. Crick and Watson had built on Oswald Avery and Erwin Chargaff's findings and elucidated the structure of the DNA molecule with a model of a double helix. See Horace F. Judson (1996) *The Eight Day of Creation*.

9 The base-pairing suggestion was made by Erwin Chargaff in the late 1940's (Chargaff 1978).

10 In 1961, J. Heinrich Matthaei and Marshall Nirenberg published their landmark paper in *Proceedings of the National Academy of Sciences*. They showed that a synthetic messenger RNA made of only uracils can direct protein synthesis. The polyU mRNA resulted in a poly-phenylalanine protein. They had the first piece of the genetic code. By 1966, Nirenberg and his group had deciphered the entire genetic code by matching amino acids to synthetic triplet nucleotides. Nirenberg and his group also showed that with few exceptions, the genetic code was universal to all life on earth. Nirenberg shared the 1968 Nobel Prize in Physiology or Medicine with Har Gobind Khorana and Robert Holley.

11 See Harold Varmus (2009) *The Art and Politics of Science*. Varmus was awarded the Nobel Prize for his discovery of the cellular origin of retroviral oncogenes.

12 Winston Churchill, speaking in Fulton, Missouri, March 5,1946: "From Stettin in the Baltic to Trieste in the Adriatic, an iron curtain has descended across the Continent. Behind that line lie all the capitals of the ancient states of Central and Eastern Europe. Warsaw, Berlin, Prague, Budapest, Belgrade, Bucharest and Sofia, all these famous cities and the populations around them lie in what I must call the Soviet sphere, and all are subject in one form or

another, not only to Soviet influence but to a very high and, in many cases, increasing measure of control from Moscow."

13 Lysenko's domination of Russian biology continued until Khrushchev's fall in 1964. Later when the noted biochemist Zhores Medvedev wrote the story of Lysenko in 1967 authorities did not react, however reprisals followed for allowing its publication in English in the United States. Medvedev lost his job and became in 1970 the first dissident intellectual to be put into an insane asylum.

14 Coming to terms with its past was rendered especially painful in view of the fact that the majority of the country's Jews had been executed in Nazi death camps. After the war, more than two million German-speaking Czechoslovak citizens were punished for their supposed Nazi sympathies. They were forced to seek a new homeland despite their roots in the Czech lands dating back to the 12th century. See Mary Heimann (2009) *Czechoslovakia*.

15 František Šorm (February 28, 1913 – November 18, 1980) was a Czech chemist known for synthesis of natural compounds, mainly terpenes and biologically active components of plants. Šorm, the founder of the Institute, studied at the Faculty of Chemistry of the Czech Technical University (later Institute of Chemical Technology, VŠCHT). After the war he returned to the university and in 1946 was named Professor at the VŠCHT. In 1950 Šorm was named Professor of organic chemistry at the Charles University in Prague. In 1952 Šorm became the Director of the Institute. During 1962–69 he served as the second President of the Academy. In the field of bioorganic chemistry, F. Šorm advanced knowledge of sesquiterpenoids, with medium-ring molecules, and explained the structure of different isoprenoid compound. See the website of the Institute of Organic Chemistry and Biochemistry, Prague (https://www.uochb.cz/web/structure/637.html?searchString=Šorm&searchId=815).

16 The first genuinely Czech national scientific institution was chartered in 1890, it was closely related to the Royal Bohemian Society of Sciences created by emperor Joseph II in 1784. After the communist party took control of the government in 1948, all scientific, non-university institutions and learned societies were dissolved to make room for the Czechoslovak Academy of Sciences. Czech chemistry was made famous by Bohuslav Brauner of the Prague University who studied rare earth elements and placed them as a separate row at the bottom of Mendeleev's periodic table. See Ulf Lagerkvist (2012) *The Periodic Table and a Missed Nobel Prize*.

17 A notable chemist from the Academy, Jaroslav Heyrovsky, received the Nobel Prize in chemistry in 1959. Heyrovsky had invented the polarographic method used in chemistry labs the world over, his Polarographic Institute was also brought under the umbrella of the Academy in 1952, see also *Nobel Lectures*, Chemistry 1942–1962 (1999).

18 See Mary Heimann (2009) *Czechoslovakia*.

19 In his *Recollections* (1994) Otto Wichterle (1913–1998) explains why he and his colleagues did not engage in the reform movement in its early phase.

20 See Jan Vilcek (2015) *Love and Science*.

21 See Zhores A. Medvedev (1969) *The Rise and Fall of T.D. Lysenko*.

22 Milan Hašek came close to the Nobel Prize. The prize was awarded instead to Peter Medawar and Burnet in 1960.

23 See Juraj Ivanyi (2003) "Milan Hašek and the discovery of immunological tolerance".

24 Quote from František Šorm, the second president of the Academy of Sciences: *The character of science is international; the results of research are in essence a synthesis of the work of scientists from all over the whole world, resulting from their close mutual cooperation or exchange of experiences. For that reason, we stand for completely free contacts of scientists in all countries and of course also that all the scientific works would be published and be public. Science is the property of the whole of humanity.* Quote researched and translated by Riika Nisonen-Trnka (2010).

25 See Horace F. Judson (1996) *The Eight Day of Creation*.

26 Interview with Prof. Marc van Montagu.

27 See Zhores A. Medvedev (1969) *The Rise and Fall of T.D. Lysenko*. The author was punished when this book was published in the United States, he was the first Russian scientist to be sent to an asylum in 1970. Fortunately many scientists from all over the world united in their protests to Soviet authorities, he was liberated and sent into exile instead.

28 Hašek would pay a heavy price as late as 1970 when he would be stripped of his membership in the party and his Directorship of his institute.

29 His connection with industry made him an attractive target for Stanford's energetic provost working to bring the best and the brightest to Silicon Valley. The provost coaxed Djerassi and his whole group of collaborators into moving to Palo Alto, as was often done with other esteemed scientists, Nobel Prize winners and their teams.

30 1964 was a landmark year. The growth of foreign academic exchanges contacts with capitalist countries, in particular in natural sciences, was extraordinary. Czechoslovakia had more extensive scientific projects going with Western countries than any of the other communist countries. The slow, inefficient intra-bloc collaboration had pushed Czechoslovakia inexorably towards more Western cooperation.

31 At first, the increase in Western contacts was depicted as a by-product of the overall improvement of international scientific cooperation. See Riikka Nisonen-Trnka (2010) "The Prague Spring of Science".

32 Otto Wichterle's licence was sold to a small American company, National Patent Development Corporation which hitherto had been focusing its interests in technology available in the Soviet Union

33 In Carl Djerassi's words "instead of poisoning the bug, one might interfere with a natural process for survival such as molting. A group in Sorm's laboratory had been working on the chemistry of insect hormones and we had initiated a collaborative research program between Zoecon and the Czech Academy [...]." See Carl Djerassi (1992) *The Pill, Pygmy Chimps, and Degas' Horse.*

34 See Antonín Holý *(2006) "My Life With Nucleic Acid Chemistry".*

Chapter III. Strange bedfellows: a Czech chemist and a Flemish virologist

1 Antonín Holý studied organic chemistry at the Charles University in Prague from 1954 to 1959.

2 See Antonín Holý *(2006) "My Life With Nucleic Acid Chemistry"*

3 Antonín Holý's assistant was Mrs. Bela Novakova.

4 Sir Alexander Todd had received a Nobelprize in 1957 for his work to trace the fundamental chemical structure of nucleic acids, the material that passes genetic characteristics from mother cell to the offspring.

5 As its new name revealed, a new department of Biochemistry was added to the Institute, the Director was Šorm's spouse, Zora.

6 See René Thomas (1992) "Molecular Genetics Under an Embryologist's Microscope".

7 Both Brachet and the Swedish scientist, Caspersson, amassed evidence for a role of RNA in "protein synthesis" in 1939 that was definitely proven about twenty years later. They were nominated for a Nobel Prize several times, with

a thorough review of their contributions by the Nobel Committee in 1959. See *Nobel Prizes and Life Sciences* by Norrby, E.

8 The Nazis imprisoned Jean Brachet together with several other professors of the Brussels University in a fortress in Southern Belgium. They all had communist ties and refused to reveal which one among their colleagues was Jewish.

9 Antonín Holý also received an honorary doctorate from the Ghent University.

10 The USA became another important scientific partner next to Germany. Scientists were discretely encouraged to emigrate to the US. See Riikka Nisonen-Trnka (2010) "The Prague Spring of Science".

11 SAFIA: Société Anonyme pour Favoriser l'Industrie Agronomique.

12 Erik and his father often walked through the neighboring village, Sint Amands, the home of the Flemish author and poet Emile Verhaeren. The French admired his mastery of the French language and wanted to immortalize Verhaeren's remains in the Pantheon in Paris. According to his wishes, Verhaeren was buried in the banks of the Scheldt so that even in death he could feel the tides of the river.

13 "Lysosomes are like the rooms within a cell" dixit Prof. Erik De Clercq.

14 The laboratory was headed by prof. Raymond Devis.

15 Steroid hormones help control metabolism, inflammation, immune functions, sexual developments, water balance and the ability to withstand illness and injury. Catecholamines are hormones produced by adrenal glands in response to physical or emotional stress.

16 An instrument used in chemistry analysis to measure light reflections.

17 By the mid-sixties geneticists could confirm what had been innate wisdom for centuries. Each time an infectious disease hit our ancestors it weeded out the weakest, leaving only the more resistant survivors to pass on their genes to future generations. Thus step by step, a long line of forebears who survived disease spawned offspring with genetic resistance to a whole range of microbes. These battles taught microbes to restrain their virulence so as not to kill their host outright while at the same time to avoid being conquered by the human's immune system. Of the million or so microbes in existence, only 1,415 are known to cause disease in humans. See Dorothy H. Crawford (2007) *Deadly Companions*.

18 Cfr. Erik De Clercq's recollections.

Chapter IV. The sixties in Leuven and Prague

1. Virus interference had first been described in 1937 by British virologist Fred Mac Callum. Virus interference as a biological phenomenon referred to the blockage by one virus of the growth of another virus when both try to infect the same cells. In the 1940s new techniques revealed that a virus could retain its capacity to 'interfere' even when it has itself been inactivated. The interference phenomenon remained nevertheless shrouded in mystery. See Toine Pieters (2005) *Interferon: Science and Selling of a Miracle Drug.*
2. Werner and Gertrude Henle studied this phenomenon in 1943 in Philadelphia. In *The Story of Interferon* Kari Cantell (1998) speculates that the Henle's could have found interferon first if they had used small bits of membrane of the fertilized chicken egg rather than whole eggs containing a fertilized embryo. In post-war England these were in short supply. Isaacs and Lindenmann substituted for the use of a whole embryonated egg embryonic membrane cultures. One single egg used sparingly would provide many pieces of the embryonic membrane. Isaacs was also in charge of the World Influenza Centre in Mill Hill, influenza viruses greatly aided his work on interferon.
3. See Derek Burke (2009) "The Discovery of Interferon, the First Cytokine, by Alick Isaacs and Jean Lindenmann in 1957".
4. See Derek Burke (2009), *ibid.*
5. One of Isaacs's letters to Lindenmann underscores this point. See Toine Pieters (2005) *Interferon: Science and Selling of a Miracle Drug.*
6. Derek Burke recalled that Isaacs received a letter from John Enders in June 1959 in a state of euphoria. See Toine Pieters (2005) *Interferon: Science and Selling of a Miracle Drug.*
7. See interview with De Somer's assistant, Edward De Maeyer, in Sandra Panem (1984) *The Interferon Crusade.*
8. An honorary doctorate five years before his untimely death, Isaacs died in January 1967.
9. See Jan Vilcek (2015) *Love and Science.*
10. Cantell's work was downplayed by those who were working with fibroblasts as they pointed out the possibility that viruses present in the donor blood might be transmitted to and cause disease with the recipients. See Kari Cantell (1998) *The Story of Interferon.*

11 When De Somer's assistant, Edward De Maeyer, tried to row upstream and publish some of his findings, he received rejection letters from both *Virology* and *Science* on the grounds that the material under study was ill-defined and impure and therefore of no great scientific interest. See Toine Pieters (2005) *Interferon: Science and Selling of a Miracle Drug.*
12 Piet De Somer *et al.* (1967) "Urinary excretion of interferon in rabbits".
13 Piet De Somer *et al.* (1968) "Antiviral activity of polyacrylic and polymethacrylic acids".
14 See Toine Pieters (2005) *Interferon: Science and Selling of a Miracle Drug.*
15 Maurice Hilleman made or improved twenty vaccines for the Merck company. See Paul A. Offit (2007) *Vaccinated.*
16 See Michael Žantovsky (2014) *Havel: A Life.*
17 The radio broadcast condemned "the occupation by the armies of the Warsaw Pact as a flagrant transgression of the principles of international law and state sovereignty which damages the cause of socialism in the eyes of all the nations of the world." See Mary Heimann (2009) *Czechoslovakia.*
18 Quote from Jaroslav Kožešník, the third president of the Academy of Sciences. See Stanley B. Winters (1994) "Věda a politika: vzestup a pád Česckoslovenské Akademie Ved [Science and Politics: The Rise and Fall of the Czechoslovak Academy of Sciences]".
19 A minor planet (3993 Šorm) was named after him in 1988 by his friends in the West. The IOCB now awards a medal named after František Šorm. Cfr website IOCB. Carl Djerassi named a sterol after Šorm, Šormosterol. See Carl Djerassi (1992) *The Pill, Pygmy Chimps, and Degas' Horse.*
20 Holý, Antonín (1967) "Synthesis of 5'deoxyuridine 5'-phosphonic acid".
21 See Stanley B. Winters (1994) "Věda a politika: vzestup a pád Česckoslovenské Akademie Ved [Science and Politics: The Rise and Fall of the Czechoslovak Academy of Sciences]".

Chapter V. Enzymes: the secret of life as chemistry

1 Erik De Clercq had an Eli Lilly fellowship for one year in Stanford and prolonged his stay for a second year with a Damon Runyon fellowship.
2 Hilleman's articles appeared in the prestigious *Proceedings of the National Academy of Sciences* in 1967. Each article revealed a different method to obtain double stranded RNA, a rare commodity. It could be found in certain molds or

in a reo virus or it could be synthesized in the laboratory imitating the chemical building blocks of RNA into poly nucleotides. (poly I:C).

3 The more common function for RNA is to read the double stranded DNA and translate it into a single strand to direct the cell in its protein production.

4 Around 1966, Merigan and De Clercq had discovered, independently from each other, that some synthetic polymers had antiviral activity.

5 Eckstein had sent copolymers that happened to be sulfur bearing RNA, compounds similar to those Maurice Hilleman had discovered.

6 See the article "Modified RNA Aids Fight against Viral Diseases" in *Chemical & Engineering* of June 1969.

7 Growing up in a poor Jewish neighborhood of New York City, Kornberg (1919–2007) gained a degree in medicine, became a ship's doctor during wartime, and had published a brief study on jaundice and vitamins or coenzymes. This caught the attention from the NIH who hired him to work on coenzyme synthesis despite his lack of formal training in science. Kornberg became the chair of the microbiology department of the University in St Louis where he found the enzymes that assemble nucleotides into RNA or DNA. This brought him to the attention of Stanford's provost who asked him to set up a new school of biochemistry on the Stanford campus. Kornberg was allowed to hire all his former co-workers and brought his team of 22 people with him to Stanford in 1959. Shortly after he started his tenure he was awarded the Nobel Prize in Medicine. See *Profiles in Science* of the NIH (https://profiles.nlm.nih.gov/).

8 See Arthur Kornberg (1989) *For the Love of Enzymes.*

9 Arthur Kornberg repeats a citation of De Clercq (1979) in his handbook *DNA Replication* (Kornberg 1992).

10 Their merger was the steppingstone for Smith Kline's expansion on the continent and later for a conglomeration with Glaxo. Today, Glaxo Smith Kline is the largest vaccine producer in the world. The vaccine based on the Cendehill strain is still the vaccine of choice in many parts of the world for MMR vaccination against rubella in combination with the measles and mumps virus vaccine. See Alfons Billiau (2011) "Piet de Somer, het Leuvense Rega Instituut en het Belgische Poliovaccin in 1956–57".

11 SKF became what is today the GlaxoSmithKline giant. SKF did not like the Rega Institute's emphasis on interferon research. See interview with Piet De Somer in *Knack Magazine*. SKF itself had abandoned its activities in this field by 1970. See Sandra Panem (1984) *The Interferon Crusade*.

12 De Clercq's technical aide was Anita van Lierde, she remained his faithful assistant till the end of her career.
13 The conference is chronicled in Kari Cantell (1998) *The Story of Interferon.*
14 Tom Merigan showed that interferon could affect the course of chronic hepatitis B infections in 1975. It opened the road for interferon to become a routine treatment for the various forms of chronic hepatitis. See Kari Cantell (1998) *The Story of Interferon.*
15 Erik De Clercq was teaching biochemistry in Kortrijk, a regional campus of the KU Leuven as of 1972. He climbed up the academic ladder and was named full Professor in 1977.
16 See the articles by Sol Spiegelman in the PNAS.

Chapter VI. From interferon to nucleosides

1 The Ghent University had conferred an honorary doctorate to David Shugar in 1969.
2 The Errera House Rue Royale, 14 Brussels, is presently the official residence of the Flemish Government.
3 The organizers were the Max Planck Institut für Biophysikalische Chemie and the Max Planck Institut für Experimentelle Medizin.
4 See photo with list of participants.
5 The first antiviral drugs developed in the late fifties and early sixties were: idoxuridine, vidarabine, trifloridine and ribavirine.
6 See Antonín Holý (2006) "My Life With Nucleic Acid Chemistry".
7 The technician was Anita Van Lierde.
8 The name, herpes, is derived from the Greek word "herpeion," meaning reptile in likely reference to the creeping nature of the herpes lesions. See Dorothy H. Crawford (2011) *Viruses.*
9 The compound, Acyclovir, was first synthesized by the US branch of the Wellcome company and its antiviral activity discovered in the UK parent company.
10 There are three herpes subfamilies; alpha, beta and gamma categorized according to the cell types in which they establish latency. So far eight human herpes viruses have been discovered named HHV 1 to 8 in order of their discovery. They also have been given common names by which they are more familiarly known, herpes simplex, varicella zoster, Epstein-Barr,

cytomegalovirus, Kaposi sarcoma associated herpes virus. See Dorothy H. Crawford (2011) *Viruses*.

11 Peter Langen was Department Head at the Institute of Biochemistry in East Germany (Berlin-Buch). An Institute that was placed under the umbrella of the Academy of Sciences of the GDR.

12 The conference was sponsored by the Federation of European Biochemistry Societies (FEBS) and co-organized with the IOCB of the Czechoslovak Academy of Sciences.

13 East Germany would become BVDU's lifeline. Finding an industrial partner in the West in order to translate the compound into a drug was fraught with misunderstandings. The drug was first produced in the early eighties in East Germany for immunosuppressed patients. BVDU—under the brand name Helpin—became available for all other patients and in the whole of Germany once the Berlin Wall fell in 1989. A comparative clinical trial in Erfürt 1995 showed that BVDU is many times more powerful than acyclovir.

Chapter VII. Breaking away from interferon

1 See Kari Cantell (1998) *The Story of Interferon.*

2 The Fifth Aharon Katzir-Katchalsky Conference: Symposium on Interferons and The Control of Cell-Virus Interactions, Rehovot, Israel, May 2–6, 1977.

3 The term, cytokine, was derived from the Greek words *kytos* for cell—inspired by the fact that these proteins are both derived from cells and act on cells—and *kine* referring to the proteins moving the immune system into action. It was not known then that hundreds of cytokines have been discovered in recent decades. See Jan Vilcek (2015) *Love and Science*.

4 Biogen was saved from bankruptcy in 1979 once it agreed to assign its future interferon patents to a pharma company, Schering-Plough. See Kari Cantell (1998) *The Story of Interferon*.

5 Maurice Hilleman wanted to find some nucleic acid or polynucleotide which could be used as a drug in humans. Poly I: C was a highly effective inducer in cell cultures and animals, but proved inactive in man because human blood contains an enzyme which breaks it down. When a stable form of poly I: C was developed, it was found that this was quite toxic to man. See Kari Cantell (1998) *The Story of Interferon*.

Chapter VIII. The first antiviral drugs

1 The NATO Advanced Study Institute in Corfu was organized by Prakash Chandra, head of molecular medicine at the University Medical School in Frankfurt.

2 Gallo's quest actually succeeded not so long after their meeting. In 1980 Gallo found a retrovirus that caused leukemia in humans he would find a second one in 1982 and unfortunately the third one, the one that causes AIDS is the one that got him embroiled as of 1983 in a Transatlantic dispute with his French colleagues.

3 Erik De Clercq found in 1975 that suramin was active against the Moloney murine leukemia retro virus. This was later seen as the first line of defense against the virus causing AIDS (Mitsuya *et al.* (1984) "Suramin protection of T cells in vitro against infectivity and cytopathic effect of HTLV-III.").

4 This NATO Advanced Study Institute took place in Sogesta, Italy and was held from 7–18 May, 1979.

5 See Richard T. Walker, Erik De Clercq and Fritz Eckstein (eds.) (1979) *Nucleoside Analogues*.

6 According to Arthur Kornberg, the di-deoxy nucleotides—much the same as the nucleotide building blocks of DNA, but lacking the chemical group that would allow them to be linked into a chain—that emerged from the work at Syntex were invaluable to two Nobel Prize winners: himself in his studies of the DNA polymerase action, which, in turn, inspired Fred Sanger to use them in his celebrated procedure for sequencing DNA.

7 John Moffatt's graduate work with Khorana dealt with the synthesis of phosphate compounds. See Arthur Kornberg (1995) *The Golden Helix*.

8 Gobind Khorana (1922–2011) had a late start in science. He grew up in the only literate family in a Punjabi village of one hundred people. Monthly visits by an itinerant teacher were hardly the preparation he needed for a university curriculum but he managed to graduate at a Punjab University and obtain a doctorate in Liverpool. It was in Cambridge with Sir Alexander Todd that he developed an interest in proteins and nucleic acids. He shared the Nobel Prize in 1968 with Marshall Nirenberg for establishing that the biological language common to all living organisms, is spelled out in three-letter words: each set of three nucleotides codes for a specific amino acid. See DNA Learning Center, Cold Spring Harbor laboratory.

9 Prusoff synthesized the first antiviral used in clinical practice, a drug called idoxuridine that treats herpes infection in the eye. It is a synthetic nucleoside.

Chapter IX. AIDS emerges in the shadow of the cold war

1 The "Charité" is an impressive 300-year old university hospital built by King Fredric II of Prussia.
2 See George Packer (2014) "The Quiet German".
3 Roald Hoffmann shared the Nobel Prize for chemistry in 1981 with a Japanese chemist, Kenichi Fukui.
4 See Hans-Jörg Schmidt (2013) "Wie Merkel Filzpantoffeln nach Prag schmuggelte".
5 See Hartmut Wewetzer (2013) "Bundeskanzlerin Merkel ehrt ihren akademischen Lehrer Rudolf Zahradník".
6 See Angela Merkel *et al.* (1988) "Evaluation of the rate constant for the SN2 reaction flouromethane+hydride → methane+fluoride in the gas phase".
7 See Hartmut Wewetzer (2013) "Bundeskanzlerin Merkel ehrt ihren akademischen Lehrer Rudolf Zahradník".
8 The Centers for Disease Control and Prevention, the leading national public health institute of the United States based in Atlanta, see their *Morbidity and Mortality Weekly Report of 5th June 1981.*
9 The documentary by Michael Isikoff (2015) *Uniquely Nasty: The U.S. Government's War on Gays.*
10 The frustration was big enough for Whitley and Martin to start thinking about creating a nucleosides network of their own. One without a limititation of American participants. This was achieved with the annual ICAR conferences of the International Society for Antiviral research (ISAR) they had founded in the mid 1980s.
11 See Sandra Panem (1988) *The AIDS Bureaucracy.*
12 See Dorothy H. Crawford (2011) *Viruses.*
13 The viruses causing leukemia, HTLV-I and HTLV-II, provoked an erratic and out-of control growth of white blood cells while the AIDS virus did quite the opposite. It attacked and destroyed white blood cells that normally protect the body's immune system.
14 See Leonard Norkin (2014) "How The Human Immunodefiency Deficiency Virus (HIV) Got Its Name" and Morton A. Meyers (2012) *Prize Fight.*

15 See Guido Van der Groen (2015) *In het spoor van Ebola.*
16 The mission to Zaire as the Democratic Republic of Congo was known then, was funded by the NIAID with a contribution from the Institute for Tropical Medicine in Antwerp.
17 It allowed the Antwerp group to set up the first AIDS project in Africa, the projet SIDA in Kinshasa.
18 Two antibody tests had been developed by 1985 that, if used together, were capable of screening out nearly all individuals infected with HIV. The first test, *enzyme-linked immunosorbent assay (ELISA) allowed for many false positives, but when followed up with a second test, the Western Blot, that showed more false negatives, they were nearly 100% accurate. See Jonathan* Engel (2006) *The epidemic.*
19 See Jonathan Engel (2006) *ibid.*

Chapter X. From passivity to action

1 Sakuma's systems to test the varicella-zoster virus and tests for the cytomegalovirus originated in the Asahikawa University on Hokkaido, Japan.
2 De Somer's speech about the "right to err," to stray from the dogmatic (Catholic) course and to change accepted standards was perceived as insubordination against the authority of the Church.
3 A Scientist at Burroughs Wellcome, Phil Furman, had found AZT's chemical structure, so the company was awarded the "method-of-use" patent.
4 Ironically, PMEA is precisely the compound that revealed its anti-HIV activity in 1988 (at the hands of Rudi Pauwels).
5 The fact that Hudson and hundreds of other Americans were seeking medical treatment outside the US—in France alone more than four hundred pleading for help—to receive Montagnier's experimental AIDS drug was a sneer to American healthcare. Secretary of Health Heckler announced the drug would be available in the US for "compassionate" use within 3 weeks.
6 Senator Jesse Helms attached anti-gay Amendments in appropriation bills one after another. The amendmends blocked prevention measures and limited research.
7 Berlin Chemie, the industrial patron of De Clercq's friends, became part of the global Italian Menarini group, based in Florence, after the Fall of the Wall. It was a new take-off for the drug and a source of royalties for the University in Leuven from 1990 to 2002. The drug was known under different names:

Helpin or Zostex in Germany, Brivirac in Italy, Zerpex in Belgium. It is now available in large parts of the world including China.

8 Blough, H.A. *et al.* (1986) "Glycosylation inhibitors block the expression of LAV/HTLV-III (HIV) glycoproteins".

9 The article was sent to Madrid in November 1986 and published in the Biochemical, Biophysical Research Communications (BBRC) of January 17, 1987. See Masanori Baba, *et al. (1987)* "Both 2′,3′-dideoxythymidine and its 2′,3′-unsaturated derivative (2′,3′-dideoxythymidinene) are potent and selective inhibitors of human immunodeficiency virus replication in vitro".

10 See William Prusoff (2001) "The Scientist's Story".

11 The final patent, use for treatment of HIV infection, was granted to Tai-shun Lin and William Prusoff in 1990. See John C. Martin *et al.* (2010) "Early nucleoside reverse transcriptase inhibitors for the treatment of HIV".

Chapter XI. First attempts to halt the epidemic

1 Janssen founded his research laboratory in 1953 within his father's company with a loan of three thousand dollars from his father. He also discovered his first drug ambucetamide that same year. In 1956 he established the company which would become Janssen Pharmaceutica. In 1958 he made haloperidol, a major breakthrough in schizophrenia and with his team he developed the fentanyl family of drugs and many other anesthesia-related drugs befor the merger with J&J in 1961.

2 See Geerdt Magiels (2008) *Paul Janssen.*

3 See Jonathan Engel (2006) *The epidemic.*

4 President Reagan and Prime Minister Jacques Chirac ended the scientific dispute between France and the U.S. They signed an agreement on March 31 about the sharing of credit for the discovery of HIV. The patent rights to a blood test that emerged from that discovery would also be shared with most of the royalties donated to a new foundation for AIDS research. The settlement contained a seven page chronology specifying the contributions of each, of Montagnier as well as of Gallo.

5 See Laurence K. Altman (1987) "U.S and France end Rift on AIDS".

6 Luc Montagnier was also awarded an honorary doctorate in 1987 by the Leuven University, on a separate occasion.

7 Galasso was the driving force behind the III international Conference on AIDS that took place a few weeks later in Washington DC, with Vice president George H. Bush as one of the main speakers. The conference assembled more than 6,000 participants.

8 Adefovir dipivoxil or Hepsera as the PMEA compound was known was the second anti-HBV drug for the treatment of chronic hepatitis B. It proved active against those HBV strains that were resistant to lamivudine, the first anti-HBV drug.

9 HPMPC consists of cytosine instead of adenine, it was first described in 1987 and approved in 1996 for retinitis in AIDS patients under the brand name Vistide.

10 Gertrude Elion had easy access for publishing through the prestigious National Academy of Sciences. She explained in her articles that acyclovir masqueraded as a nucleoside, a building block of DNA, in order to be incorporated into the herpes virus DNA and prevent it from further using the cell's machinery to replicate. (Gertrude Elion *et al.* (1977), "Selectivity of action of an antiherpetic agent, 9- (2-hydroxyethoxymethyl)guanine").

11 In 1988, Gertrude Elion of Burroughs Wellcome (now GlaxoSmithKline) received the Nobel Prize in Medicine along with George Hitchings and Sir James Black for "their discoveries of important principles in drug treatment." Gertrude Elion's name was indelibly associated with Acyclovir.

Chapter XII. Finding the best therapy: the one-a-day-pill

1 Paula Pitha had earned her Ph.D under the directorship of František Šorm at the IOCB in Prague. She hosted an event for Erik De Clercq's lecture at Johns Hopkins in 1970, after he left Stanford University. Cf. chapter V.

2 The company was incorporated under a place-holder name and then officially changed to Gilead Sciences a few months later, in early 1988. Michael Riordan came across "Gilead" when he read a play by the American playwright Lanford Wilson, called "Balm in Gilead." He found that the balm is an extract from certain trees indigenous in a region of the ancient Middle East called Gilead, near the Jordan River. The balm of Gilead was probably one of mankind's first genuine therapeutics. Riordan added "Sciences" to the name to make clear the company was based on rigorous research and scientific principles.

3 A former Gilead scientist, Jeff Bird, spoke with reporters, about the early days of Gilead. See Denise Gellene "Gilead's Research Goes to Front Lines".

4 Benno C. Schmidt Sr. (1913 - 1999) was an American lawyer who is credited with coining the term venture capital. See *Richard A. Oppel Jr (1999) "Benno C. Schmidt, Financier, Is Dead at 86".*

5 Etienne (Stevie) Davignon, was a former Belgian diplomat and former Member of the European Commission. The common bond between Rumsfeld and Davignon was Stevie's mentor, the legendary Belgian ambassador at NATO and one time confidant of Winston Churchill, André de Staercke. He had joined Rumsfeld's board at Searle but was rapidly advancing in age and deferred to his friend, Davignon.

6 Gilead's Initial public offering was of 5.75 million common shares of stock. The IPO was completed, with proceeds of $86,25 million.

7 Erik De Clercq had convinced Paul Janssen to travel to Oxnard, California to participate in the session devoted to "the non-nucleoside anti-HIV drugs." The topic was close to his heart, since they were precisely his compounds that Rudi Pauwels and Dominique Schols had been so doggedly working with in the Rega Institute.

8 Gancyclovir was still not approved by FDA but it had saved many people already who were on the drug in parallel circuits.

9 Robert Snoeck had done all the screenings and testings in the clinic. Another of De Clercq's coworkers, an Argentinian researcher, Graciela Andrei, found many other properties of cidofovir including one to treat cancer caused by HPV. De Clercq met Graciela Andrei at a conference in Mar del Plata and enticed her to work with him. She had received some special training at the University of Alabama to work with the human papilloma viruses, a very large family of viruses with over 100 different types, a few types can cause cancer but can be treated in the clinic with *cidofovir*.

10 Compounds from the IOCB in Prague were handled in the Rega Institute by Jan Balzarini, Dominique Schols and Lieve Naessens.

11 The three protease inhibitors were produced by Merck (which made Crixivan), Abbott (Norvir), and Hoffman-La Roche (Invirase).

12 The center was created by Irene Diamond in memory of her husband who left her several hundred millions of dollars upon his death, Irene Diamond was famous for her production of the movie "Casablanca." See Jonathan Engel (2006) *The epidemic*.

13 Viread and Emtriva were marketed as single drugs and patients were required to take other Antiretrovirals. Similarly patients taking Truvada had to also take either a Non-Nucleoside Reverse Transcriptase Inhibitor (NNRTI) or a protease inhibitor (PI).

Epilogue: Of scientists and crusaders

1 Peter Piot, em. professor at the Institute of Tropical Medicine in Antwerp, he became Assistant Director of the World Health Organization's Global Programme on HIV/AIDS. He was appointed Exceutive Director of the Joint United Nations Programme–UNAIDS and Assistant-Secretary-General of the United Nations (1994–2008).

2 The Security Council adopted Resolution1308 in July 2000, stating that there should be no UN peacekeeping operations without HIV prevention.

3 See Peter Piot (2015) *AIDS Between Science and Politics.*

4 See Joe Conason (2016) *Man of the World:. The Further Endeavors of Bill Clinton.*

5 The World Health Organization estimates that HIV has claimed more than 35 million lives so far and that approximately 36.7 million people were living with HIV at the end of 2015. By mid-2016, 18.2 million people living with HIV were receiving antiretroviral therapy. See Media centre HIV/AIDS of the World Health Organization.

References

Altman, Lawrence K. (1987), "U.S. and France End Rifts on AIDS", *New York Times Special* April 1, 1987.

Atlas, Ronald M. (ed.) (2000) *Many faces, Many Microbes. Personal Reflections in Microbiology* (ASM Press)

Baba, Masanori, Rudi Pauwels, Piet Herdewijn, Erik De Clercq, Jan Desmyter, and Michel Vandeputte (1987) "Both 2′,3′-dideoxythymidine and its 2′,3′-unsaturated derivative (2′,3′-dideoxythymidinene) are potent and selective inhibitors of human immunodeficiency virus replication in vitro" *Biochemical and Biophysical Research Communications*, 142(1):128–134.

Billiau, Alfons (2009a) "A brief history of interferon's trajectory to clinical application, and personal reminiscences of a large-scale human interferon production intiative", *Verhandelingen van de Koninklijke Academie voor Geneeskunde van België*, 71(1–2): 15–42.

Billiau, Alfons (2009b) "Penicilline in België", *Verhandelingen van de Koninklijke Academie voor Geneeskunde van België*, 71(4):165–203.

Billiau, Alfons (2011) "Piet de Somer, het Leuvense Rega Instituut en het Belgische Poliovaccin in 1956–57", *Verhandelingen van de Koninklijke Academie voor Geneeskunde van België*, 73(3–4):189–250.

Billiau, Alfons (2014) "A Polio Vaccine for Belgium in 1956", *Quarterly Newsletter of the Belgian Society for Microbiology* 8: 7–11.

Booss, John and Marilyn J. August (2013) *To Catch a Virus* (ASM Press).

Blough, H.A., R. Pauwels, E. De Clercq, J. Cogniaux, S. Sprecher-Goldberger, and L. Thiry (1986) "Glycosylation inhibitors block the expression of LAV/HTLV-III (HIV) glycoproteins" *Biochem. Biophys. Res. Commun.* 141(1): 33–38

Burke, Derek (2009) "The Discovery of Interferon, the First Cytokine, by Alick Isaacs and Jean Lindenmann in 1957", *Brain Immune Trends* February 14, retrieved from http://brainimmune.com/the-discovery-of-interferon-the-first-cytokine-by-alick-isaacs-and-jean-lindenmann-in-1957/

Brachet, Lise (2004) *Le professeur Jean Brachet, mon père* : biologiste moléculaire (L'Harmattan).

Cantell, Kari (1998) *The Story of Interferon. The Ups and Downs in the Life of a Scientist* (World Scientific Publishing).

Chargaff, Erwin (1978) *Heraclitean Fire: Sketches from a Life before Nature* (The Rockefeller University Press).

Conason, Joe (2016) *Man of the World: The Further Endeavors of Bill Clinton* (Simon and Shuster).

Crawford, Dorothy H. (2007) *Deadly Companions: How microbes shaped our history* (Oxford University Press).

Crawford, Dorothy H. (2011) *Viruses: A Very Short Introduction* (Oxford University Press).

De Kruif, Paul (1926 &1996) *Microbe Hunters* (Harcourt, Inc.).

Djerassi, Carl (1992) *The Pill, Pygmy Chimps, and Degas' Horse* (Harper Collins)

De Clercq, Erik, E. Eckstein and T.C. Merigan (1969) "Interferon induction increased through chemical modification of a synthetic polyribonucleotide" *Science* 165(3898): 1137–9.

De Clercq, Erik and Antonín Holý (1979), "Antiviral activity of aliphatic nucleoside analogs: structure-function relationship" *Journal of Medical Chemistry* 22: 510–513.

De Clercq, Erik (ed.) (1987) *Frontiers in Microbiology* (Martinus Nijhoff)

De Clercq, Erik (1982) "Interferon: a Molecule for all Seasons" in *Virus Infections. Modern Concepts and Status,* ed. Lloyd C. Olson (Marcel Dekker) 87–138.

De Clercq, Erik (2015) "An Odyssey in antiviral drug development: 50 years at the Rega Institute: 1964–2014" *Acta Pharmaceutica Sinica* B (5(6):520–543.

de Duve, Christian (2004) "The first milligrams of Belgian penicillin" speech 50th anniversary of the Rega Institute, October 9, 2004.

De Somer Piet, Erik De Clercq, Alfons Billiau, E. Schonne (1967) "Urinary excretion of interferon in rabbits" *Proceedings of the First International Conference on Vaccines against Viral and Rickettsial Diseases of Man*, Pan American Health Organisation, Fort Lauderdale, Fl, USA, 1967, 650–652.

De Somer Piet, Erik De Clercq, Alfons Billiau, E. Schonne and M. Claesen (1968) "Antiviral activity of polyacrylic and polymethacrylic acids", *Journal of Virology*, 2(9):878–885.

Elion, Gertrude B., Phillip A. Furman, James A. Fyffe, Paulo de Miranda, Lilia Beauchamp and Howard J. Schaeffer (1977), "Selectivity of action of an antiherpetic agent, 9- (2-hydroxyethoxymethyl)guanine", *Proceedings of the National Academy of Sciences* USA, 74(12): 5716–5720.

Engel, Jonathan (2006) *The epidemic: A Global History of AIDS* (Smithsonian Books).

France, David (2016) *How to Survive a Plague. The Inside Story of How Citizens and Science Tamed AIDS* (Alfred A. Knopf).

Field, Dawn and Neil Davies (2015) *Biocode: The New Age Of Genomics* (Oxford University Press)

Fox Renée C. (1994) *In the Belgian Château: The Spirit and Culture of a European Society in an Age of Change* (Ivan R. Dee).

Fox, Renée C. (1997) *Experiment Perilous: Physicians and Patients Facing the Unknown* (Routledge).

Garfield, Eugene (1992) "The Restoration of František Šorm: Prolific Czech Scientist Obeyed His Conscience and Became a Nonperson" *Essays of an Information Scientist: Of Nobel Class, Women in Science, Citation Classics and Other Essays*, 15, 75 (*Current Comments*, 23, June 8, 1992, Institute for Scientific Information, Philadelphia).

Garrett, Laurie (1994) *The Coming Plague: Newly Emerging Diseases in a World out of Balance* (Penguin books).

Gallo, Robert C. (1991) *Virus Hunting: AIDS, Cancer and the Human Retrovirus* (Basic Books).

Gellene, Denise (2003) "Gilead's Research Goes to Front Lines", *The New York Times*, July 31.

Hargittai, István (2002) *The Road to Stockholm. Nobel Prizes, Science, and Scientists* (Oxford University Press).

Halleux, Robert, Carmélia Opsomer and Jan Vandersmissen (1998) *Geschiedenis van de wetenschappen in België van de Oudheid tot 1815* (Gemeentekrediet)

Heimann, Mary (2009) *Czechoslovakia: The State that Failed* (Yale University Press).

Holý, Antonín (1967) "Synthesis of 5'deoxyuridine 5'-phosphonic acid" *Tetrahedon Letters* 8(10):881–884.

Holý, Antonín (2006) "My Life With Nucleic Acid Chemistry", *Collection of Czechoslovak Chemical Communications* 71(8):v-xi.

Ivanyi, Juraj (2003) "Milan Hašek and the discovery of immunological tolerance" *Nature Reviews Immunology* 3(7):591–7.

Judson, Horace F. (1996) *The Eight Day of Creation: Makers of the Revolution in Biology* (Cold Spring Harbor Laboratory Press).

Kornberg, Arthur (1989) *For the Love of Enzymes: The Odyssey of a Biochemist* (Harvard University Press).

Kornberg, Arthur (1992) *DNA Replication* (University Science Books) 447.

Kornberg, Arthur (1995) *The Golden Helix: Inside Biotech Ventures* (University Science books).

Knight, Amy (2005) *How the Cold War Began: The Igor Gouzenko Affair and the Hunt For Soviet Spies* (Caroll & Graf publishers).

Lagerkvist, Ulf (2003) *DNA Pioneers and Their Legacy* (World Scientific Publishing).

Lagerkvist, Ulf (2012) *The Periodic Table and a Missed Nobel Prize* (World Scientific Publishing).

Levere, Trevor H. (2001) *Transforming Matter: A History of Chemistry from Alchemy to the Buckyball* (Johns Hopkins University Press).

Martin, John C., Michael J.M. Hitchcock, Erik De Clercq and William H. Prussoff (2010) "Early nucleoside reverse transcriptase inhibitors for the treatment of HIV: A brief history of stavudine (D4T) and its comparison with other dideoxynucleosides" *Antiviral Research* 85(1):34–8.

Magiels, Geerdt (2008) *Paul Janssen: Pioneer in Pharma and in China* (Dundee University Press).

"Modified RNA AIDS fight against viral diseases" Chem Eng News 1969, 23 June:17–8.

Morrison, John (2005) *Mathilde Krim and the Story of AIDS* (Chelsea House Publishers).

McMullen, Chris (2012) *Understanding Basic Chemistry Concepts* (Northwestern State University of Louisiana)

Medvedev, Zhores A. (1969) *The Rise and Fall of T.D. Lysenko* (Columbia University Press)

Merkel, Angela, Zdenek Havlas and Rudolf Zahradnik (1988) "Evaluation of the rate constant for the SN2 reaction flouromethane+hydride.fwdarw. methane+fluoride in the gas phase" *Journal of the American Chemical Society* 110(25):8355–8359.

Meyers, Morton A. (2012) *Prize Fight: The Race and Rivalry to be the First in Science* (Palgrave Macmillan).

Mitsuya, H., M. Popovic, R. Yarchoan, S. Matsushita, R.C. Gallo and S. Broder (1984) "Suramin protection of T cells in vitro against infectivity and cytopathic effect of HTLV-III", *Science*, 226(4671):172–4.

Nagorski, Andrew (1993) *The Birth of Freedom: Shaping Lives and Societies in the New Eastern Europe* (Simon & Schuster).

Matthaei, J. Heinrich and Marshall Nirenberg (1961) "Characteristics and stabilization of DNA ase-sensitive protein synthesis in E. coli extracts" *Proceedings of the National Academy of Sciences* 47(10):1580–1588.

Nisonen-Trnka, Riikka (2010) "The Prague Spring of Science: Czechoslovak Natural Scientists Reconsidering the Iron Curtain", in *1948 and 1968 – Dramatic Milestones in Czech and Slovak History*, ed. Laura Cashman (Routledge) 105–122.

Leonard Norkin (2014) "How The Human Immunodefiency Deficiency Virus (HIV) Got Its Name" *Virology*, February 4, 2014 (retrieved from https://norkinvirology.wordpress.com/2014/02/04/how-the-human-immunodeficiency-deficiency-virus-hiv-got-its-name/)

Nobel Lectures, Chemistry 1942–1962 (1999, World Scientific).

Offit, Paul A. (2007) *Vaccinated: One Man's Quest to Defeat the World's Deadliest Diseases* (Harper Collins).

Oppel Richard A. Jr. (1999) "Benno C. Schmidt, Financier, Is Dead at 86", *The New York Times*, October 22, 1999.

Oshinsky, David M. (2005) *Polio: An American Story. The Crusade That Mobilized the Nation Against the 20th Century's Most Feared Disease* (Oxford University Press).

Pačes, Václav (2014) "Biochemistry Behind The Iron Curtain", in *FEBS at 50: Half a century promoting the molecular life sciences*, eds. Mary Purton and Richard Perham (Third Millennium Publishing) 120–122.

Packer, George (2014) "The Quiet German: The astonishing rise of Angela Merkel, the most powerful woman in the world", *The New Yorker*, December 1, 2014.

Panem, Sandra (1984) *The Interferon Crusade* (The Brookings Institution).

Panem, Sandra (1988) *The AIDS Bureaucracy: Why Society Failed to Meet the AIDS Crisis and How We Might Improve Our Response* (Harvard University Press).

Pieters, Toine (2005) *Interferon: Science and Selling of a Miracle Drug* (Routledge).

Piot, Peter (2015) *AIDS Between Science and Politics* (Columbia University Press).

Pringle, Peter (2012) *Experiment Eleven: Dark Secrets Behind the Discovery of a Wonder Drug* (Bloomsbury)

Prusoff, William (2001) "The Scientist's Story", *The New York Times*, March 19, 2001.

Richter, Jan (2012) interview with Rudolf Zahradník, *Radio Praha* 03-04-2012.

Riedel, Stefan (2005) "Edward Jenner (1749–1823) and the history of smallpox and vaccination" *Baylor University Medical Center Proceedings* 18(1):21–25

Rumsfeld, Donald (2011) *Known and Unknown: A Memoir* (The Penguin group).

Schmidt, Hans-Jörg (2013) "Wie Merkel Filzpantoffeln nach Prag schmuggelte", *Die Welt*, December 18, 2013.

Teitelman, Robert (1989) *Gene Dreams: Wall Street, Academia, and the Rise of Biotechnology* (Basic Books/Harper Collins).

Thomas, René (1992) "Molecular Genetics Under an Embryologist's Microscope: Jean Brachet, 1909–1988" *Genetics* 131(3):515–518.

Van der Groen, Guido (2015) *In het spoor van Ebola. Mijn leven als virusjager* (Lannoo).

Varmus, Harold (2009) *The Art and Politics of Science* (W.W. Norton).

Vilcek, Jan (2015) *Love and Science: a Memoir* (Seven Stories Press).

Walker, Richard T., Erik De Clercq and Fritz Eckstein (eds.) (1979), *Nucleoside Analogues: Chemistry, Biology and Medical Applications* (Plenum Press).

Waterson, A. P. and Lise Wilkinson (1978) *An Introduction to the History of Virology* (Cambridge University Press).

Watson, J.D. and F.H.C. Crick (1953) "Molecular Structure of Nucleic Acids: A structure for Deoxyribose Nucleic Acid" *Nature* 171:737–738.

Wewetzer, Hartmut (2013) "Bundeskanzlerin Merkel ehrt ihren akademischen Lehrer Rudolf Zahradník", *Der Tagesspiegel*, November 3, 2013.

Wichterle, Otto (1994), *Recollections* (Evropský Kulturní Klub).

Winters, Stanley B. (1994) "Věda a politika: vzestup a pád Česckoslovenské Akademie Ved [Science and Politics: The Rise and Fall of the Czechoslovak Academy of Sciences]" *Bohemia, Zeitschrift für Geschichte und Kultur der Böhmischen Länder*, 35(2):286–287.

Žantovsky, Michael (2014) *Havel: A Life* (Atlantic Books).

Websites

www.scienceguardian.com

www.nobelprize.org

www.History.org

www.Erikdeclercq.org

http://www.ornl.gov/hgmis (U.S. Department of Energy Human Genome Project)

https://profiles.nlm.nih.gov

Index

Lightning Source UK Ltd.
Milton Keynes UK
UKOW01f1020280917
310039UK00003B/521/P

9 789462 701137